WHAT IS A MATHEMATICAL CONCEPT?

Responding to widespread interest within cultural studies and social inquiry, this book addresses the question of what a mathematical concept is by using a variety of vanguard theories in the humanities and post-humanities. Tapping into historical, philosophical, sociological and psychological perspectives, each chapter explores the question of how mathematics comes to matter. Of interest to scholars across the usual disciplinary divides, this book tracks mathematics as a cultural activity, drawing connections with empirical practice. Unlike other books in this area, it is highly interdisciplinary, devoted to exploring the ontology of mathematics as it plays out in different contexts. This book will appeal to scholars who are interested in particular mathematical habits – creative diagramming, structural mappings, material agency, interdisciplinary coverings – that shed light on both mathematics and other disciplines. Chapters are also relevant to social sciences and humanities scholars, as each one offers philosophical insight into mathematics and how we might live mathematically.

Elizabeth de Freitas is a professor in the Education and Social Research Institute at Manchester Metropolitan University. Her research focuses on philosophical investigations of mathematics, science and technology, and pursuing the implications and applications of this work within cultural studies. She is a co-author of the book *Mathematics and the Body: Material Entanglements in the Classroom*, associate editor of the journal *Educational Studies in Mathematics* and has written more than 50 chapters and articles on diverse topics.

Nathalie Sinclair is the Canada Research Chair in Tangible Mathematics Learning at Simon Fraser University. She is the author of several books, including co-author of *Mathematics and the Body: Material Entanglements in the Classroom* and co-editor of *Mathematics and the Aesthetic: New Approaches to an Ancient Affinity*, as well as more than 50 articles. She has also led the design of educational technologies, including the touchscreen app TouchCounts and dynamic geometry microworlds for young learners (www.sfu.ca/geometry4yl). She is the founding editor of the journal *Digital Experiences in Mathematics Education*.

Alf Coles' recently published *Engaging in School Mathematics: Symbols and Experiences* draws on more than 20 years of work as a teacher-researcher at both primary and secondary levels. He is on the executive committee of the British Society for Research into Learning Mathematics and is an active member of the Mathematics Education Special Interest Group of the British Educational Research Association. His current interests include drawing his work in mathematics education into closer dialogue with issues of sustainability.

What Is a Mathematical Concept?

Edited by

Elizabeth de Freitas

Manchester Metropolitan University

Nathalie Sinclair

Simon Fraser University

Alf Coles

University of Bristol

CAMBRIDGE
UNIVERSITY PRESS

CAMBRIDGE
UNIVERSITY PRESS

One Liberty Plaza, 20th Floor, New York, NY 10006, USA

Cambridge University Press is part of the University of Cambridge.

It furthers the University's mission by disseminating knowledge in the pursuit of education, learning, and research at the highest international levels of excellence.

www.cambridge.org
Information on this title: www.cambridge.org/9781107134638
DOI: 10.1017/9781316471128

First published 2017

A catalogue record for this publication is available from the British Library.

Library of Congress Cataloging-in-Publication Data
Names: De Freitas, Elizabeth. | Sinclair, Nathalie. | Coles, Alf.
Title: What is a mathematical concept? / [edited by] Elizabeth de Freitas,
Manchester Metropolitan University, Nathalie Sinclair, Simon Fraser University,
Alf Coles, University of Bristol.
Description: Cambridge: Cambridge University Press, 2017. |
Includes bibliographical references and index.
Identifiers: LCCN 2016059487 | ISBN 9781107134638 (hard back)
Subjects: LCSH: Mathematics – Social aspects. | Mathematics – Philosophy.
Classification: LCC QA10.7.W43 2017 | DDC 510.1–dc23
LC record available at https://lccn.loc.gov/2016059487

ISBN 978-1-107-13463-8 Hardback

CONTENTS

IMAGES

NOTES ON CONTRIBUTORS

YASMINE ABTAHI is a part-time professor at the Faculty of Education, University of Ottawa and post-doctoral research fellow at the Université du Québec à Montréal. Her research includes work on mathematical tools and artefacts.

MICHAEL J. BARANY is a postdoctoral fellow in the Dartmouth College Society of Fellows. He recently completed his PhD in Princeton University's Program in History of Science with a dissertation on the globalization of mathematics as an elite scholarly discipline in the mid-twentieth century. His research on the relationship between abstract knowledge and the modern world has led to articles (all available at http://mbarany.com) on such topics as dots, numbers, rigour, blackboards, basalt, bureaucracy, communism and internationalism, from the sixteenth century to the present.

RICHARD BARWELL is Professor of Mathematics Education at the Faculty of Education, University of Ottawa. His research includes work on language, multilingualism and discourse analysis in mathematics education. He was educated in the United Kingdom before moving to Canada in 2006. Prior to his academic career, he taught mathematics in the United Kingdom and Pakistan.

TONY BROWN is Professor of Mathematics Education at Manchester Metropolitan University, where he also leads research in teacher education. Brown's work explores how contemporary theory provides new insights into educational contexts. He has written seven books including three volumes for Springer's prestigious Mathematics Education Library series. He convenes the Manchester-based conference on Mathematics Education and Contemporary Theory.

ALF COLES is Senior Lecturer in Education (Mathematics) at the University of Bristol. He gained a research council scholarship for his PhD study that was adapted as a book: *Being Alongside: For the Teaching and Learning of Mathematics* (2013). His research covers early number development, creativity in learning mathematics, working on video with teachers and links between mathematics education and sustainability education. His latest book, *Engaging in School Mathematics* was published by Routledge in 2015.

DAVID CORFIELD is Senior Lecturer in Philosophy at the University of Kent. He works in the philosophy of science and mathematics and is a co-director of the Centre for Reasoning at Kent. He is one of the three owners of the blog The n-category Café, where the implications for philosophy, mathematics and physics of the new language of higher-dimensional category theory are discussed. In 2007, Corfield published *Why Do People Get Ill?* (co-authored with Darian Leader), which aims to revive interest in the psychosomatic approach to medicine.

BRENT DAVIS is Professor and Distinguished Research Chair in Mathematics Education in the Faculty of Education at the University of Calgary. He is the author of two books on pedagogy and co-author of three books on learning, teaching and research. He has served as editor of *For the Learning of Mathematics* (2008–2010), co-editor of *JCT: Journal of Curriculum Theorizing* (1995–1999), and founding co-editor of *Complicity: An International Journal of Complexity and Education* (2004–2007).

SIMON B. DUFFY received a PhD in Philosophy from the University of Sydney in 2003 after a Diplôme d'Etudes Approfondies (MPhil equivalent) in Philosophy from the Université de Paris X-Nanterre (1999). He has taught at the University of Sydney, the University of New South Wales and the University of Queensland, where he was a postdoctoral fellow in Philosophy at the Centre for the History of European Discourses. Dr Duffy is the author of *Deleuze and the History of Mathematics: In Defense of the New* (2013) and *The Logic of Expression: Quality, Quantity and Intensity in Spinoza, Hegel and Deleuze* (2006). He is editor of *Virtual Mathematics: The Logic of Difference* (2006), and co-editor with Sean Bowden of *Badiou and Philosophy* (2012). He is also translator of Albert Lautman's *Mathematics, Ideas and the Physical Real* (2011).

ELIZABETH DE FREITAS is a professor at the Education and Social Research Institute at Manchester Metropolitan University. She is the co-author of *Mathematics and the Body: Material Entanglements in the Classroom* (Cambridge University Press, 2014) and *Alternative Theoretical Frameworks for Mathematics Education Research: Theory meets Data* (2016). Her work focuses on the philosophy and history of mathematics and its implications for theories of learning and pedagogy. She is an associate editor of the journal *Educational Studies in Mathematics*.

MICHAEL HARRIS is a professor of mathematics at the Université de Paris Diderot and Columbia University. He is the author or co-author of more than seventy mathematical books and articles and has received a number of prizes, including the Clay Research Award, which he shared in 2007 with Richard Taylor. His most recent book is *Mathematics without Apologies: Portrait of a Problematic Vocation* (2014).

JULIETTE KENNEDY is Professor of Mathematics at the University of Helsinki. Her research interests include set theory and set-theoretic model theory, foundations and philosophy of mathematics, history of logic and aesthetics and art history. She has published several books including, most recently, *Interpreting Gödel: Critical Essays* (2014). She also co-organised the Simplicity, Ideals of Practice in Mathematics and the Arts conference in New York.

HEATHER MENDICK is a sociologist and a former mathematics teacher who currently works as a freelance academic. She is the author of *Masculinities in Mathematics* (2006), the co-author of *Urban Youth and Schooling* (2010) and the co-editor of *Mathematical Relationships in Education* (2009) and *Debates in Mathematics Education* (2014). Her most recent research project focused on the role of celebrity in young people's classed and gendered aspirations and was funded by the Economic and Social Research Council (www.celebyouth.org). She tweets about work, politics, darts and pop culture @helensclegel.

RICARDO NEMIROVSKY is Professor at Manchester Metropolitan University and a faculty member of the Education and Social Research Institute. Dr. Nemirovsky's research focuses on informal STEM education, museum pedagogy and embodied cognition. He has acted as PI on a number of National Science Foundation grants, including projects focusing on art- science museum collaborations. He has designed numerous interactive tools and manipulatives for mathematics learning and is the author of many seminal articles pertaining to mathematics and cognition, such as the co-authored Mathematical Imagination and Embodied Cognition (2009).

REVIEL NETZ is the Patrick Suppes Professor of Greek Mathematics and Astronomy at the Department of Classics, Stanford University. He has written widely on Greek mathematics, and among his books are *The Shaping of Deduction in Greek Mathematics: A Study in Cognitive History* (Cambridge University Press, 1999) and *The Archimedes Palimpsest* (co-edited with W. Noel., 2011).

DAVID PIMM is Professor Emeritus at the University of Alberta. He is the author of *Speaking Mathematically* (1987) and *Symbols and Meanings in School Mathematics* (1995) and a co-author of *Developing Essential Understanding of Geometry* (2012). He is a former editor of the journal *For the Learning of Mathematics* (1997–2003) and has written extensively on mathematics and mathematics education, drawing on both the history and the philosophy of mathematics.

ARKADY PLOTNITSKY is Professor of English and Theory and Cultural Studies, director of the Theory and Cultural Studies Program and co-director of the Philosophy and Literature Program at Purdue University. He earned his PhD in comparative literature and literary theory from the University of Pennsylvania and his MSc in mathematics from the Leningrad (St. Petersburg) State University in Russia. He has published several books including *Niels Bohr and Complementarity: An Introduction* (2012), *Epistemology and Probability: Bohr, Heisenberg, Schrödinger, and the Nature of Quantum-Theoretical Thinking* (2009) and *Complementarity: Anti-Epistemology after Bohr and Derrida* (1994).

WOLFF-MICHAEL ROTH is Lansdowne Professor of Applied Cognitive Science at the University of Victoria. He conducts research on how people across their lifespan know and learn mathematics and science. He is a Fellow of the American Association for the Advancement of Science, the American Educational Research Association (AERA) and the British Society. He received a Significant Contribution award from AERA and an Honorary Doctorate from the University of Ioannina, Greece.

NATHALIE SINCLAIR is a professor in the Faculty of Education, an associate member in the Department of Mathematics and the Canada Research Chair in Tangible Mathematics Learning at Simon Fraser University. She is also the editor of *Digital Experiences in Mathematics Education*. She is the author of *Mathematics and Beauty: Aesthetic Approaches to Teaching Children* (2006), and co-author of *Mathematics and the Body: Material Entanglements in the Classroom* (Cambridge University Press, 2014) and *Developing Essential Understanding of Geometry* (2012), among other books.

Introduction

Responding to widespread interest within cultural studies and social inquiry, this book takes up the question of what a mathematical concept is, using a variety of vanguard theories in the humanities and posthumanities. Tapping into historical, philosophical, mathematical, sociological and psychological perspectives, each chapter explores the question of how mathematics comes to matter. Of interest to scholars across the usual disciplinary divides, this book tracks mathematics as a cultural and material activity. Unlike other books in this area, this book is highly interdisciplinary, devoted to exploring the ontology of mathematics as it plays out in empirical contexts, offering readers a diverse set of crisp and concise chapters.

The framing of the titular question is meant to be simple and direct, but each chapter unpacks this question in various ways, modifying or altering it as need be. Authors develop such variations as:

1. When does a mathematical concept become a mathematical concept?
2. What is the relationship between mathematical concepts, discourse and the material world?
3. How might alternative ontologies of mathematics be at work at this historical moment?
4. How do our theories of cognition and learning convey particular assumptions about the nature of mathematical concepts?
5. How might we theorize processes of mathematical abstraction and formalisation?
6. What is the role of diagrams, symbols and gestures in making mathematical concepts?
7. How do mathematical concepts inform particular ideological positions?

The authors take up these questions using tools from philosophy, anthropology, sociology, history, discursive psychology and other fields, provoking readers to interrogate their assumptions about the nature of mathematical concepts. Thus, the book presents a balance of chapters, diverse in their application but unified in their aim of exploring the central question. Each chapter examines in some detail case studies and examples, be they historical or situated in contemporary practice and public life. Each author explores the historical and situated ways that mathematical concepts come to be valued. Such focus allows for a powerful investigation into how mathematical concepts operate on various material planes, making the book an important contribution to recent debates about the nature of mathematics, cognition and learning theory. In offering a set of diverse and operational approaches to rethinking the nature of mathematics, we hope that this book will have far-reaching impact across the social sciences and the humanities. Authors delve into particular mathematical habits – creative diagramming, tracking invariants, structural mappings, material agency, interdisciplinary coverings – in order to explore the many different ways that mathematical concepts come to populate our world.

THE CONTEXT FOR THIS BOOK: PHILOSOPHY AND COGNITION

This book springs from our desire to pursue a cultural studies of mathematics that incorporates philosophy, history, sociology, and learning theory. We conceived this book as a collection of essays exploring and in some sense reclaiming a canonical question – what is a mathematical concept? – from the philosophy of mathematics. Authors take up this question innovatively, tapping into new theory to examine contemporary mathematics and current contexts. For those unfamiliar with the philosophy of mathematics, this section briefly recounts how this canonical question was typically addressed in the past. The ontology of concepts has long been a central concern for philosophers, and many of these philosophers considered the mathematical concept as an exemplary case for their investigations. The conventional starting point has tended to be framed as a dichotomy: Do mathematical concepts exist inside or outside the mind? From this starting point, further binaries are encountered: If concepts exist outside the mind, are they corporeal or incorporeal? If they are corporeal, do they exist in the things that are perceptible by the senses or are they separate (or independent) from them? Bostock (2009) suggests that philosophers have typically taken three positions in relation to such questions: cognitive, realist and nominalist.[1] These conventional responses have dominated the philosophy

[1] We have changed Bostock's term "conceptualist" to "cognitive" better to name its focus on *mental* concepts, and to avoid any confusion with how the term is used in our book.

of mathematics in previous centuries, and have become somewhat ossified in their characterization. This book charts entirely new territory, and yet for the sake of context it is worth describing very briefly these three schools of thought, and tracing their influence on twentieth-century constructivist theories of learning. This will set the stage for the post-constructivist approaches that are used in this book.

The cognitive approach claims that concepts exist in the mind and are created by the mind. Descartes, Locke and Kant, to some degree, might be considered to be in this camp. According to some variants of the cognitive approach, humans create universal, matter-independent concepts based on sense perception, while other variants claim that concepts are innate and do not require perceptual experience. In either case, concepts are treated as mental images or language-like entities. The second group of Bostock's philosophers, the realists (e.g., Plato, Frege and Gödel), claim that mathematical concepts exist outside the mind and are independent of all human thought, while the third group, the nominalists, claim that they do not exist at all, and are simply symbols or fictions.

Of course such sorting of philosophers into simplistic positions ignores the complexity of their thought, but it might help some readers, who are unfamiliar with the philosophy of mathematics, appreciate the radically divergent approaches developed in this book. Moreover, it is important to note how particular ideas from this tradition – such as Kant's theory that mathematical statements are "synthetic *a priori*" – have saturated many later developments in the philosophy of mathematics, seeping into the realist and nominalist camps as well. Brown (2008) indicates that Frege embraced Kant's view on geometry, Hilbert embraced Kant's view on arithmetic and even Russell can be characterized as Kantian in some crucial respects.

One might also argue that Kant's theory of mathematical truth has saturated theories of learning and has become full fledged in cognitive psychology and its dominant image of learning as that which entails acquiring a set of cognitive 'schemas'. Constructivist theories of learning, in which concepts are constructed rather than acquired, also tend to frame the constructed concept as a mental image. According to this approach, student capacity for developing mathematical concepts is based in part on inductively generalising from engagements with material objects and discourse. A constructivist approach to concept formation tends to centre on the epistemic subject who synthesizes and subsumes these diverse materials and social encounters under one *cognitive* concept. Accordingly, concepts are treated as abstractions that ultimately transcend the messy world of hands, eyes, matter and others.

Constructivist theories of concept formation find their usual source in the work of either Piaget or Vygotsky. In the former case, Piaget's notion of *reflective abstraction* has been used to describe what it means to learn or develop a concept. Piaget spoke of four different types of abstractions, but the notion of reflective abstraction that was adopted by many education researchers involves the dual process of projection (borrowing existing knowledge from a preceding level of thought to use at a higher level) and conscious reorganization of thought into a new structure (becoming aware of what has been abstracted in that projection). For Piaget, reflective abstraction was the mechanism through which all mathematical structures were constructed. In his genetic epistemology approach, he broke with existing theories of concept development found both in philosophy and psychology because he based his analyses on empirical observations of children's activity. For example, in the case of number, Piaget combined the relational and classificatory concepts of number, which had been seen as incommensurable by philosophers at the time (Brainerd, 1979). This focus on the mathematical activity of non-experts introduced important insights that philosophers had overlooked. On the other hand, researchers today who follow in the Piagetian tradition (see, for example, Simon et al., 2016) tend to pay little attention to philosophical considerations of particular mathematical concepts, focusing exclusively on the trajectories of particular children working on particular tasks.

For Vygotsky, concept formation was goal-oriented and entirely social: "A concept emerges and takes shape in the course of a complex operation aimed at the solution of some problem" (1934, p. 54); "A concept is not an isolated, ossified, and changeless formation" (Vygotsky, p. 98). Vygotsky saw concept formation as necessarily being mediated by signs (principally language and material tools); for instance, he argued that language is the means by which a learner focuses attention and makes distinctions within the environment, distinctions that can be analysed and synthesized. As with Piaget, Vygotsky insisted that concepts could not be taught directly, and that concept formation was a long and complex process. Whereas spontaneous concepts could be developed from direct experience of the world through induction, scientific concepts develop through deduction and require exposure (through school, for example) to abstract cultural knowledge and different forms of reasoning. Thus, one way of characterizing the difference between Piaget and Vygotsky is that for the former, reflective abstractions begin with the actions of the individual and are then shared out in the social realm, while for the latter, scientific concepts begin in the social realm and are internalized by the individual. Researchers working through a Vygotskian perspective today focus strongly on the role that language and tools play in learners' concept formation, as well as on the teacher actions that support the process of internalization (see, for example, Mariotti, 2013).

The tendency for researchers influenced by both Piaget and Vygotsky to focus almost exclusively on the psychological nature of concepts may account for DiSessa and Sherin's (1998) critique of current educational work on concepts. In their attempt to formalise "conceptual change", they note that one of the main difficulties in most accounts is "the failure to unpack what 'the very concepts' are in sufficiently rigorous terms" (p. 1158). This frustration might stem in part from the fact that researchers cannot *see* the schemes or structures that are posited by Piaget's account of reflective abstraction, or even the process of internalisation described by Vygotsky.

In the context of education research, concepts are often distinguished from memorized facts and procedures, and often qualified in terms of misconceptions and protoconceptions. Curriculum policy advocates for the importance of conceptual understanding, and typically stipulates which mathematical concepts are most important in teaching and learning. But this kind of listing of key concepts offers little insight into the specific nature of mathematical concepts and the material-historical processes associated with them.

Recent developments in post-constructivist learning theories have shown how concepts are performed, enacted or produced in gestures and other material activities (Davis, 2008; Hall & Nemirovsky, 2011; Radford, 2003; Roth, 2010). This new theoretical shift draws attention to how concepts are formed in the activity itself rather than in the rational cognitive act of synthesizing (Brown, 2011; Tall, 2011). This work reflects a paradigmatic shift in learning theory, driven in large part by offshoots of contemporary phenomenology, better to address the role of the body in coming to know mathematics.

There are yet further developments on this front, developments that build on the phenomenological tradition, and diverge from it in significant ways. For instance, Deleuze and Guattari (1994), whose work is cited often in this book, reanimate the concept as part of their philosophy of immanence. They propose a "pedagogy of the concept", by which concepts are to be treated as creative devices for carving up matter, rather than pure forms subject only to recognition. This pedagogy of the concept aims to encounter and engage with the conceptual on the material plane; a concept brings with it an entire "plane of immanence" (Cutler & MacKenzie, 2011, p. 64). For Stengers (2005), Deleuze's pedagogy is about learning "the 'taste' of concepts, being modified by the encounter with concepts" (p. 162). de Freitas and Sinclair (2014) have developed this post-humanist approach to concept formation, arguing that learning is about encountering the mobility and indeterminacy of concepts.

This book takes up these recent developments to explore new ontologies of mathematics and pushes against all-too-easy dualisms between matter and meaning. It does so by taking a broad view of concepts to include their historical and cultural dimensions, their trajectories in and through classrooms and their potentially changing nature within contemporary mathematics. The chapters dig deep into mathematical practice and culture, troubling conventional approaches and their constructivist offspring.

Our hope is that this book contributes to the philosophy of mathematics (how does mathematics evolve as a discipline? How are concepts formed and shared?), as well as cultural studies of mathematics (How do mathematical concepts format worldviews? How do they participate in the creation of political and social discourse?). We also hope that the book triggers discussions about significant questions within mathematics education, such as: How might learning theories change if we view concepts as generative of new spacetime configurations rather than timeless, determinate and immovable? What happens to curriculum when we treat concepts as material assemblages, temporally evolving and vibrating with potentiality?

THEMES AND CHAPTERS

The first two chapters are by Michael J. Barany and Reviel Netz, respectively, who each provide some more historical context (and critique) of theories of mathematical concept construction. Barany engages in some long-standing considerations of the epistemological status of mathematical concepts, with a particular interest in the principle of *meaning finitism*, which emerged from sociology of scientific knowledge (SSK) perspectives that gained currency in the 1970s. This perspective stresses the contingent human aspects of mathematical knowledge, particularly through the activities of labelling and classifying. Barany uses Lakatos' account of the development of the concept of polyhedron to exemplify a "meaning finitism" account of mathematics. Rather than focus on more ontological debates about the status of simple objects (numbers, shapes), Barany focuses on how mathematical concepts are used and revised over time.

Netz's chapter raises the question of what it means for mathematics to be conceptual, especially in the context of historical situations. He describes many claims that have been made about whether or not certain cultures possessed a particular mathematical concept. He highlights two ways in which such claims might be misleading. The first relates to what we might call frequency of use. Netz shows several examples of a concept existing in a certain culture without it becoming widespread or frequently used. The second, perhaps more interesting to mathematicians, relates to conceptual hierarchy. By showing persuasively how Archimedes used the concept of actual infinity, Netz troubles common assumptions that the concept of actual infinity depends on the concept of set. As Barany's meaning finitism would make evident, the particular ways in which knowledge is classified (ordered, related) is highly contingent and cannot be assumed to play out in the same way in different historical periods and different geographical locations. Indeed, Netz highlights how different mathematical *practices* give rise to different concepts.

The next two chapters continue to look at the material practices of mathematical activity, exploring how mathematical concepts live through various media. Juliette Kennedy examines the role of visualization and diagramming in mathematics, and asks whether some mathematical concepts are irreducibly visual. She focuses on the role of these informal "co-exact" characteristics of mathematical drawing for the part they play in logical inference, first tracking the historical separation of the visual from the logical. The chapter by Elizabeth de Freitas and Nathalie Sinclair attends to the historical division between logic and mathematics in a related way, looking at the concept of the mathematical continuum, to show that number and line are mathematical concepts which are the source of persistent philosophical questions about space, time and mobility. Just as Kennedy talks about the "bidirectionality" of mathematical practice (between body and symbol) and the "ambivalence" entailed in mathematical positioning, de Freitas and Sinclair suggest that mathematical concepts are always rumbling beneath the apparent foundations of mathematical truth. They draw on the ideas of Gilles Châtelet and Ian Hacking to show how concepts thrive through material media and historical material arrangements. These two chapters challenge readers to reconsider the way that proof and reasoning is at play in mathematics.

Kennedy first distinguishes between drawings that are directly constitutive of a mathematical proof and others that are informal, "incidental" aspects of mathematical activity, discussing how both kinds function fruitfully in mathematics. She discusses "world-involving inference" and logical inference, seeking a middle synthetic ground where mixtures of reasoning operate. Drawing on the reflections of the architect Juhani Pallasmaa about "the thinking hand", Kennedy argues that the manual activity of mathematical drawing must be considered as we ask the question: What is a mathematical concept? Mathematicians move around a mathematical diagram much like one might move around a building, and it is through this habitation and spatial practice that the concepts become known. This chapter also links to that by Nemirovsky, who describes how one comes to inhabit a concept over time, through habitual carving out of its contour and meaning.

The chapter by de Freitas and Sinclair continues the theme that Kennedy opens, regarding the relationship between the logical and the mathematical. They cite Hacking (2014), who argues that the connection between symbolic logic and mathematics "simply did not exist" until the logicist movement of the nineteenth century (advocated by Frege, above all), which aimed to reduce mathematics to logic, and replaced Aristotelian logic with what was termed "symbolic logic" (p. 137). This chapter proposes the term "virtual" to describe the indeterminate dimension in matter that literally destabilizes the rigidity of extension. They suggest that concepts such as line, point and circle can be conceived using a genetic definition that emphasizes the dynamic and mobile aspects of mathematical concepts. Concepts – such as squareness, fiveness, etc. – thus retain the trace of the movement of the eye, hand and thinking body. This chapter is linked to the one by Netz, as they both present images of mathematical practice as an applied or practical affair, grounded in material conditions and experiments rather than exclusive appeals to logic.

Chapters by Arkady Plotnitsky and Simon Duffy explore the ways in which mathematical concepts spring from and sustain rich problem spaces. They both draw on the powerful ideas of Gilles Deleuze and Felix Guattari to develop a theory of mathematical concepts, and then show its relevance to other discourses. Deleuze, in particular, offered deep insights into the history of mathematics, tapping particular ideas – from Galois, Riemann, Poincaré, Lautman and others – to rethink the relationship between concepts and problems. We see in Plotnitsky and Duffy's chapters a theoretical move that explores the speculative position of a "*mathesis universalis*" (Deleuze, 1994, p. 181), but not one that posits a definite system of mathematical laws at the base of nature. Rather, these two chapters delve into the mathematical concept as that which operates through a rich dynamic ontology of problems that are in some way shared with other discourses and contexts.

Plotnitsky explores the contributions of Bernhardt Riemann around non-Euclidean geometry, also drawing on the insights of Deleuze. Riemann's work is known as a conceptual rather than axiomatic approach to exploring non-Euclidean geometries. Plotnitsky uses the work of Riemann to show that a mathematical concept (1) emerges from the co-operative confrontation between mathematical thought and chaos; (2) is multi-component; (3) is related to or is a problem; and (4) has a history. Plotnitsky argues that mathematical concepts are not simply referents or functional objects, but that they tap into a "plane of immanence", which is a Deleuzian term that describes the vibrant virtual realm of potentiality *in* the world. The plane of immanence is the plane of the movement of *philosophical* thought that gives rise to *philosophical* concepts, but Plotnitsky argues that mathematics also creatively operates through this plane of immanence. In particular, Plotnitsky shows how mathematical concerns regarding the distinction between discrete and continuous manifolds are philosophical in the Deleuzian sense. Thus, Plotnitsky shows that mathematics as much as philosophy engages with "chaos" by creating planes of immanence and concepts. He argues that creative *exact* mathematical and scientific thought is defined by planes of immanence and invention of *exact* concepts, the architecture of which is analogous to that of philosophical concepts in Deleuze and Guattari's sense.

Duffy shows how a *practice of mathematical problems* – using the examples of the problem of solving the quintic and the problem of the diagrammatic representation of essential singularities – operates as the engine of mathematical invention, such that the emergent "solutions" are clusters of concepts that carry with them the problem space from which they emerged. In other words, following Lautman, concepts are inherently problematic and carry with them the force of the problem – indeed, this force animates them. Duffy shows how Deleuze is ultimately interested in how this theory of mathematical problems offers even broader significance because it can be deployed *as a way of studying problems and concepts in other discourses, or fields and contexts.* In particular, Duffy shows how Deleuze's work in his seminal *Difference and repetition* (1994) deploys the conceptual space of the early mathematical calculus to rethink the nature of perception. It is not, however, that Deleuze privileges the discourse of mathematics over others in some absolute sense, but rather that it offers distinctive insights (just as any other might) into our shared ontology.

The chapters by David Corfield and Michael Harris both consider the emergence of new concepts in mathematics, in a contemporary setting. Corfield's chapter is concerned with homotopy type theory while Harris traces the recent emergence of the perfectoid. Corfield's interest in homotopy type theory stems from the way it exemplifies the vertical unity of mathematics. For Harris, the focus is on how the concept of the perfectoid came to be seen as "the right" concept within the mathematics community – a story he offers as a participant-observer. Both authors highlight how mathematical concepts are tied up in axiological concerns. While

Harris refuses to offer criteria for what makes a concept "good", he draws attention to the many social and historical factors – such as the connection to Grothendieck and perhaps even the endearing personality of Scholze – that converged to make the perfectoid the 'right' concept for solving a set of diverse mathematical problems. He chronicles the way in which the perfectoid concept was put to work extensively by Scholze and others, almost like a kind of mutant offspring of current theories. This suggests that the applicability of a concept (where the application is across mathematics, rather than outside of mathematics), is a highly generative process whereby new practices emerge that change the entire field.

Similarly, Corfield provides a compelling argument for the "goodness" of homotopy type theory, which has developed a strong footing in the past decade. Corfield describes how this theory, and type theory more generally, exploits the vertical unity of mathematics. Such unity entails consistency demands, but perhaps also points to uncharted pedagogical terrain. There are some important nuances to keep in mind, which Corfield highlights in his discussion of Mark Wilson's insistence on the "wandering" nature of concepts and his warning that "hazy holism" can often misleadingly lead us to believe in the unity of concepts, which are more often than not "patched together from varied parts" (p. 129). The very practice of *patching* becomes pivotal to Corfield's considerations of the 'spatial' nature of homotopy type theory.

Thus we might also see the vertical unity as arising from a patching together of different kinds of mathematical practices, much as we saw in Harris' chapter. That strong analogies can be seen across basic arithmetic and homotopy is convincingly and carefully shown by Corfield, but one look at the syntactic complexity required to "express" addition or inverse in homotopy type theory is enough to remind us that these are not the same concepts. We are reminded of Thurston's (1994) description of the different ways of thinking about the derivative. While the differences may "start to evaporate as soon as the mental concepts are translated into precise, formal and explicit definitions" (p. 3), they are much more real in the particular contexts in which they are actually used. Staying close to particular practices – rather than erasing those differences within a reductive set theory – allows Corfield to seek out other important "unities" across other concepts, such as formal and concrete duality.

The notion of vertical unity seems to us an interesting one for mathematics education, for how it troubles conventions about developmental conceptual change and curriculum. School mathematics has long been considered an edifice whose stairway must be climbed one step at a time. Vertical unity brings about some different imagery: express elevators, the possibility of starting at the penthouse of homotopy type theory, a confused and more wandering landscape of conceptual life. This chapter links to that of Reviel Netz, which also troubles conventional assumptions about which concepts come 'before' others.

Three chapters (Yasmine Abtahi and Richard Barwell; Tony Brown; Heather Mendick) delve into the public culture of mathematical concepts, in different ways, by tracking the way that concepts are used outside of academic mathematics. Mendick takes a post-modern perspective to ask what mathematical concepts do in popular culture ('queering' mathematical concepts in the process) in order to ask what they might do differently in the future. The chapter takes the form of a mathematical-concept archive. In one part of the archive, Mendick looks at school student Twitter responses to a recent examination in the United Kingdom to illustrate their refusal to dissociate mathematical concepts from the contexts in which questions are placed or posed. Mendick ends her chapter with a series of dichotomies that she has hoped to disrupt and that include one highlighted by Barwell and Abtabi, but which are here expressed as conceptual understanding vs procedural understanding. Both chapters are therefore working against binaries that can appear natural and inevitable but are never innocent in that they come to be used to separate, evaluate and segregate groups of students.

Barwell and Abtahi investigate how the word "concept" is used in relation to Canadian media reports about mathematics education, drawing on a corpus of 53 articles. The side-stepping of ontological questions about how concepts are coupled to the material world is a deliberate position arising from the discursive psychology perspective of their work. They discover that the phrase 'mathematical concept' gets associated with 'discovery learning', framing concept acquisition as hard or difficult, compared with the simplicity of back-to-basics routines. Discovery learning and concepts are associated with approaches to teaching that are less successful when contrasted to what are characterised as older and simpler procedure-based methods. The newspaper reports therefore set up a dichotomy with concepts (discovery learning, difficulty, confusion) on one side and routines (back-to-basics teaching, simplicity, clarity) on the other.

The ideological implications of the binaries that inform our thinking and action is taken up in the chapter by Brown, who investigates the production and commodification of mathematical concepts. Ideology is at play in, for example, the determination of mathematical truth and in the assessment structures that surround the production of school mathematics. We do not usually notice the ideology at play in relation to mathematical concepts and yet, Brown argues, mathematics often figures prominently in the making of our very subjectivity. Brown draws on the insights of Jacques Lacan to argue that through participating in rituals (for example, the ritual of school mathematics assessment) we inadvertently materialise our own belief in the ideological state apparatus associated with those mathematical rituals. And yet we can still encounter spaces 'beyond', as Brown contends that mathematical thought will always exceed its commodified manifestations, perhaps echoing the way that students in Mendick's chapter refuse to allow their encounter with mathematics to be stripped of context.

Chapters by Alf Coles and Wolff-Michael Roth, in different ways and with differing emphases, deal with a paradox of learning that has been recognised since antiquity. Plato (Meno, 80d) asks: if learning is the recognition of the new, how is this ever possible, since to recognise something I need to know what I am looking for? In a modern take, Anna Sfard referred to essentially the same paradox: to participate in a discourse on an object, you need to have already constructed this object, but the only way to construct an object is to participate in the discourse about it (see Sfard, 2013). Roth expresses the paradox in language linked to his background in cultural, historical activity theory: "*[b]ut how would an individual, who does not already know what is cultural about objects encountered sequentially come to abstract precisely those features that make some of the objects members of a cultural concept while excluding others?*" (p. 223, italics in the original). Coles cites visual theory to pose the paradox in relation to perception, suggesting we need abstract structures to make sense of perception, and we need perception to build abstract structures.

Coles also uses the word "abstract" in his framing, but in his chapter the term is taken to mean attention to relations, rather than attention to objects. However, Coles suggests it is probable that any relation can be seen as an object and any object seen as a relation – we can therefore become aware of our choice (although typically we do not notice) in engaging in object-oriented or relational thinking. Coles suggests learning mathematics can become fast, imaginative and engaging if we introduce concepts, from the very beginning, as relations. Four examples are given of how this could be done, with the most detailed example being early number – and the suggestion is made that curriculum could be taught in a relational manner.

Although Roth uses the word "abstract" in posing the paradox, his solution is decidedly concrete. Roth puts forward the "documentary method" as a way of explaining how we come to create new distinctions and new categories, and he exemplifies this with an empirical classroom example. According to the documentary method, we learn concepts that allow us to make distinctions in the world, without necessarily needing to uncover "common properties". A concept remains forever a class of concrete manifestations. With familiarity we are able to identify class membership "at a glance", but this does not mean we have erased all distinctive features from all instances of that class. Classes of objects become one if they are treated as such.

The next two chapters, by Brent Davis and Ricardo Nemirovsky, both provide expansive views of mathematical concept, seeking in some ways to free it from its static straightjacket. Davis draws on the historical connection between ideas and species – which were often seen as synonymous in pre-evolutionary science – to investigate how mathematical concepts might be studied through the lens of contemporary biology, where species are contingent, situated and volatile. Davis asks: What if concepts were seen to be more like species? He uses explorations into mimetics, complexity science and embodied cognition to propose that concepts are "memeplexes", with a life form and a networked living body that evolves in complex ways. In the context of mathematics education, Davis suggests that embracing concepts as species may compel a different attitude towards student understanding and teacher knowledge. If a concept is a living form, it makes little sense to speak of "acquiring" it; instead, Davis invites us to consider how students and teachers might be seen as propagators of ideas.

Davis' inquiry into species, and especially the associations suggested by embodied cognition in which bodies are the media of concepts, segues nicely into Nemirovsky's chapter, which considers the more anthropological perspective of *inhabiting* mathematical concepts. Nemirovsky begins by evoking the classical, arboreal image of the Aristotelian tree diagram of nodes in which concepts are seen as classes of entities (humans, triangles). He critiques this image on two grounds: its static presentation of concept as fixed within the tree, and its failure to account for the cultural and political forces that create the differences out of which the nodes are arranged. Like Davis, Nemirovsky draws on images from biology – of growth and decay – to re-imagine the mathematical concept. He describes the way concepts might be seen to grow and decay through affect and the virtual, both of which can be seen as exceeding any fixed, intrinsic determination. He exemplifies this process through the concept of number, with a particular focus on Cantor's work and its reception by Frege. Nemirovsky shows how the historical development of transfinite numbers altered the way we inhabit the concept of number. Sidestepping the usual discovery/invention debate, he suggests that "inhabiting" captures the experience of working with mathematical concepts.

In the final chapter, David Pimm provides a commentary of sorts, reading across the chapters and highlighting some notable themes. He reflects on various linguistic features evoked in relation to mathematical concepts, both in terms of how concepts are named and renamed, as well as in relation to their potential metaphorical, poetic and diagrammatic qualities. Drawing on an eclectic range of sources (such as poetry, philosophy and psychoanalysis), he then offers seventeen evocative assertions about concepts that play off particular passages found in the preceding chapters.

ON READING THE BOOK

When we first planned this book, we hoped to be able to suggest multiple pathways through the book, inspired by Julio Cortazar's *Hopscotch*. In the end, we paired up the chapters (and in one case, tripled them up) based on their tangled threaded ideas, but instead of naming the groups according to a theme, we decided to offer images for each group – which were kindly provided by friends and artists – that captured something about the duo or trio of chapters. We hope that the images work generatively, perhaps leading the reader to create connections of their own, both within each duo/trio and across the whole book. We thank David Pimm for so expertly and creatively offering his own set of connections, in the afterword that follows all the chapters, which may incite some readers to consume the book in an order different than the one we have offered.

We close by thanking the contributors and attendees of the American Education Research Association roundtable, where the idea for this book was born in 2014. We would also like to thank the Coles family (Niki, Iona, Arthur and Iris) for hosting us over a long weekend in Bristol as we gathered these chapters together and wrote this introduction.

REFERENCES

Bostock, D. (2009). *The philosophy of mathematics: An introduction*. New York: Wiley Blackwell.
Brainerd, C. (1979). *The origins of the number concept*. New York: Praeger Publishers.
Brown, J. R. (2008). *Philosophy of mathematics: A contemporary introduction to the world of proofs and pictures* (2nd Ed.). New York: Routledge.
Brown, L. (2011). What is a concept? *For the Learning of Mathematics*, 31(2), 15–17.
Cutler, A., & MacKenzie, I. (2011). Bodies of learning. In L. Guillaume & J. Hughes (Eds.), *Deleuze and the body* (pp. 53–72). Edinburgh: Edinburgh University Press.
Davis, B. (2008). Is 1 a prime number? Developing teacher knowledge through concept study. *Mathematics Teaching in the Middle School (NCTM)*, 14(2), 86–91.
de Freitas, E., & Sinclair, N. (2014). *Mathematics and the body: Material entanglements in the classroom*. New York: Cambridge University Press.
Deleuze, G. (1994). *Difference and repetition*, trans. P. Patton. London: Athlone.

Deleuze, G., & Guattari, F. (1994). *What is philosophy?* London: Verso.
DiSessa, A., & Sherin, B. (1998). What changes in conceptual change? *International Journal of Science Education*, 20(10), 1155–1191.
Hacking, I. (2014). *Why is there philosophy of mathematics at all?* New York: Cambridge University Press.
Hall, R., & Nemirovsky, R. (2011). Histories of modal engagement with mathematical concepts: A theory memo. Accessed December 2, 2016, at http://www.sci.sdsu.edu/tlcm/all-articles/Histories_of_modal_engagement_with_mathematical_concepts.pdf
Mariotti, M. A. (2013). Introducing students to geometric theorems: how the teacher can exploit the semiotic potential of a DGS. *ZDM Mathematics Education*, 45, 441–452.
Piaget, J. (1953). *The origin of intelligence in the child.* London: Routledge and Kegan Paul.
Radford, L. (2003). Gestures, speech, and the sprouting of signs: A semiotic- cultural approach to students' types of generalization. *Mathematical Thinking and Learning*, 5(1), 37–70.
Roth, W.-M. (2010). Incarnation: Radicalizing the embodiment of mathematics. *For the Learning of Mathematics*, 30(2), 8–17.
Sfard, A. (2013). Discursive research in mathematics education: Conceptual and methodological issues. In A. Lindmeier & A. Heinze (Eds.), *Proceedings of the 37th annual conference of the International Group for the Psychology of Mathematics Education* (Vol. 1, pp. 157–161). Kiel, Germany: PME 37.
Shapere, D. (1987). Method in the philosophy of science and epistemology. In Nancy J. Nersessian (Ed.), *The process of science: Contemporary philosophical approaches to understanding scientific practice* (pp. 1–39). Dordrecht, Boston, Lancaster: Martinus Hijhoff Publishers.
Simon, M. Placa, N., & Avitzur, A. (2016). Paticipatory and anticipatory stages of mathematical concept learning: Further empirical and theoretical development. *Journal for Research in Mathematics Education*, 47(1), 63–93.
Stengers, I. (2005). Deleuze and Guattari's last enigmatic message. *Angelaki*, 10(2), 151–167.
Tall, D. (2011). Crystalline concepts in long-term mathematical invention and discovery. *For the Learning of Mathematics*, 31(1), 3–8.
Thurston, W. (1994). On proof and progress in mathematics. *Bulletin of the American Mathematical Society*, 30(2), 161–177.
Vygotsky, L. S. (1962 [1934]). *Thought and language.* Cambridge, MA: MIT Press.

PART I

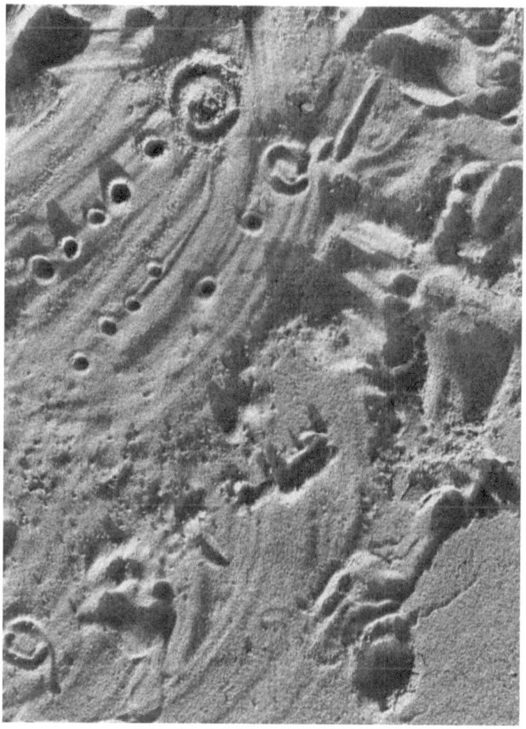

1

Of Polyhedra and Pyjamas: Platonism and Induction in Meaning-Finitist Mathematics

MICHAEL J. BARANY

INTRODUCTION

> Think not of a word being uttered, vibrating through the air, and being lost, but of the farmer clipping the sheep's ear, the nurseryman tagging the plant, the hospital issuing pyjamas. (Barnes, 1983, p. 528)

Near the start of his seminal essay on "bootstrapped induction," Barry Barnes asks the reader to think of labelling as an active enterprise, one which makes "an enduring change to the situation" (ibid.). Labels, he explains, are not passive, fleeting descriptions. Rather, they stay with their subjects like pyjamas on hospital patients, situating them in patterns of language and cognition. These "pattern attachment systems" are what enable the objects of the world to become concepts that are circulated in a social field of knowledge. Whatever objective order they may independently possess, objects cannot be ordered in the conceptual universes of the people who study them without such systems of labels.

Though mathematical concepts can be inspired by the things of the natural world and can be the basis for interventions in the world, mathematical concepts cannot bleat in a field or be arrayed in greenhouse rows. The things of mathematics appear purely through human activity, manifested in definitions, textbooks, diagrams, gestures, and utterances by and for people. Though few would suppose that mathematical concepts have the kind of existence of a tree in a forest, which one supposes could grow, photosynthesize, and fall (perhaps even making a sound) without any human intervention, there are nonetheless features of mathematical experience that suggest mathematical concepts, once formulated, do take on a certain life of their own or might even have been "out there" waiting to be formulated.[1]

This peculiar ontological character attributed to mathematical concepts raises a host of epistemological questions that are related, but not typically reducible, to corresponding questions about scientific and social knowledge. One important perspective on such questions has been the strong program of the Edinburgh School of the sociology of scientific knowledge (SSK), associated with two of its chief proponents, Barry Barnes and David Bloor. SSK debuted in the 1970s and 1980s as an exciting and sometimes controversial approach to the study of science in relativist terms, with roots in the philosophy of Ludwig Wittgenstein. It featured significantly in the "science wars" of the 1990s, and has since waned in prominence, though it continues to figure in academic curricula, research, and debates in the sociology of science.

Emphasizing the social life of labels, advocates of the strong program in SSK stress the irreducibly contingent human aspects of scientific and mathematical knowledge. By casting knowledge as fundamentally relational and revisable, the strong program's adherents instructively contrast their accounts with those who place greater weight on the supposed power of mathematical entities and inferences to compel assent in and of themselves. That is, where many philosophers aim to account for the appearance and implications of determinacy and certainty in mathematics, SSK guides one instead to emphasize the sources of *indeterminacy* in mathematical understanding and how those indeterminacies are socially resolved.

In this chapter, I place classical accounts of mathematical concepts in the context of the Edinburgh School's social epistemology, exploring (in Barnes' metaphor) what it means for a polyhedron to be issued pyjamas. I begin by sketching the principles of meaning finitism in the strong program, and then illustrate the meaning finitist account of mathematical objects by considering Imre Lakatos' (1976) depiction of Euler's theorem concerning polyhedra. Using this framework, I address in turn the relationships among meaning finitism, epistemic induction, and mathematical Platonism, which imply a special importance for simple, workable models

[1] This idea forms a central theme of Hacking (2014), a distinctive recent entry in an enormous body of scholarship on this question. Hacking argues, in part, that these features of mathematical experience are not as representative of what mathematicians spend their time doing as philosophers frequently suppose. However, the many applications of mathematical reasoning and representation in other areas of mathematics, science, and beyond still seem to demand a philosophical understanding of mathematical ideas as less-than-arbitrary.

and examples. While meaning-finitism is foremost a principle about labelling and classification, I explain how a meaning finitist account of these activities applies more broadly to the use and revision of mathematical concepts.

I conclude by indicating how Barnes' pyjama metaphor can be extended profitably to account for the process, less considered in SSK accounts of mathematics, by which simple mathematical concepts are used to establish more complex ones. This view expands the SSK account of meaning finitism from a theory about the conditions of stabilization for simple mathematical concepts into a theory that also encompasses their conditions of change. Barnes' pyjama metaphor, I argue, impels us to examine the figurative seams of mathematical practice, how concepts are negotiated and stabilized. It is not enough to describe exemplary mathematical concepts as wearing pyjamas without also asking how those pyjamas are issued, made to fit, and occasionally stripped in the process of mathematical knowledge-making.

STRONG PROGRAM MEANING FINITISM

Meaning finitism, within the strong program of SSK, is based on the idea that "concept application is a matter of judgement at the individual level, of agreement at the level of the community" (Barnes, 1982, p. 30). This entails five central claims, summarized by Barnes, Bloor, and Henry (BBH, 1996, pp. 55–59):

1. The future applications of terms are open-ended.
2. No act of classification is ever indefeasibly correct.
3. All acts of classification are revisable.
4. Successive applications of a kind term are not independent.
5. The applications of different kind terms are not independent of each other.

The first three claims describe knowledge as *flexible* and the final two emphasize that it is also *relational*. The flexibility of knowledge comes from its fundamental formal indeterminacy: every thing has similarities and differences with every other thing or collection of things, so in no case do our past classifications logically and indefeasibly compel us in our future ones (BBH, 1996, pp. 51, 78; Barnes, 1982, pp. 28–30). Some classifications are more defensible than others, but it is always possible to imagine that some (perhaps bizarre) method of identifying and weighing similarities and differences could overturn even the most obvious-seeming divisions between natural objects. Meaning finitists reject the notion that there are pre-given natural partitions of the world according to different kinds (Barnes, 1981, p. 315).

Instead, classification must proceed by "analogy between the finite number of our existing examples of things and the indefinite number of things we shall encounter in the future" (BBH, 1996, p. 51; see also Barnes, 1982, p. 49; Bloor, 1983, p. 95). Old classifications form the basis of new ones according to their degree of similarity to the objects in question, something which is always negotiable but never purely arbitrary (Barnes, 1981, pp. 309, 312; Barnes, 1982, p. 29; Bloor, 1997, pp. 10, 70). The context-dependent negotiations over classification are the basis for the social study of meaning (Barnes, 1981, p. 314; Barnes, 1982, p. 30; BBH, 1996, p. 79).

In the terms of Barnes' metaphor, a particular set of pyjamas can fit people of a variety of shapes and sizes, and an individual person can fit (to varying degrees of comfort) in a few different sizes and styles of pyjamas. If you are currently wearing hospital pyjamas, you are more likely to be issued them again in the future, but you are not consigned to a life of loose-fitting hospital garb. When a new patient arrives at the hospital, that patient's similarity to other patients can guide which pyjamas are issued, but there may be multiple sets that fit. Pyjamas, meanwhile, collapse distinctions between those wearing them, transforming distinct individuals into common patients (or patients of a certain size or shape). They make patients more uniform by temporarily covering some of their features, and by equipping them with a shared and recognizable sartorial marker.

Mathematics has a special place in the strong program for both epistemic and historical reasons. Epistemically, it differs from the natural sciences in treating what is supposedly a realm of pure ideas. This requires a reinterpretation of ostention, the ability to name a thing by gesturing at it, which has a foundational status in meaning finitist epistemology (Barnes, 1981, pp. 306, 308; Barnes, 1982, p. 35). In classical meaning finitism, the thing you are classifying is itself something at which you can point. But in mathematics, anything at which one can physically point is already at least a step removed from the mathematical ideals whose classification is at issue. To what exactly, in mathematics, would pyjamas be issued?

Historically, mathematics has represented an ideal of knowledge independent of messy human contingencies. Paradoxically, it has been imagined to be free of *both* nature *and* culture. This presents a double challenge to the fundamental approaches of SSK, which stress the interlocking roles of nature and culture in human knowledge. For this reason, Bloor (1976, p. 73) called mathematics "the most stubborn of all obstacles to the sociology of knowledge … the holy of holies."

Thus, taking a note from the later aphorisms of Ludwig Wittgenstein, strong program scholars began by attacking the self-evidence and inevitability of some of the most natural-seeming mathematical practices – including counting by twos and taking twice two to equal four (e.g. Bloor, 1973, 1983, 1997; BBH, 1996).[2] Somewhat less discussed but nonetheless canonical, Lakatos' (1976) heuristic history of Euler's theorem offered the strong program's expositors a chance to extend the case for meaning finitism to less elementary mathematics (e.g. Bloor, 1976, pp. 130–137; Bloor, 1978, pp. 248–250; BBH, 1996, p. 187). According to Euler's theorem, the number of vertices plus faces minus edges of any polyhedron is equal to two. Lakatos showed how, starting at the turn of the nineteenth century, attempts to prove this theorem involved confrontations over the very meaning of the term "polyhedron," among several contested mathematical concepts.

In Bloor's gloss on Lakatos' story, Euler's theorem began as an observation about the relationship among vertices, edges, and faces in a certain limited collection of polyhedra. This observation was generalised by Euler into the equation $V-E+F=2$, an example of epistemic induction (Bloor, 1976, p. 135). That is, based on a limited collection of observations, Euler stipulated that one could expect his formula to hold whenever one counted the vertices, edges, and faces of a polyhedron in the future. Lakatos' story unfolds in a series of historically proposed counterexamples – situations in which the expectation that the formula would hold did not seem to bear out – each of which complicates a different aspect of proofs of the theorem or definitions of polyhedra (ibid.: pp. 133–134). Because, as Bloor asserts, "Polyhedra have no essence," these counterexamples are accepted or rejected according to social interests, and the decisions "will reveal what types of figure and what features of figures are held to be important and interesting" for different mathematicians (ibid.: pp. 135–136; see also Bloor, 1978; Barnes, 1981, p. 325).

While mathematical concepts and proofs are often treated as though they are stable and unambiguous, Bloor takes Lakatos to show that "informal thought can always outwit formal thought" (Bloor, 1976, p. 137). As Wittgenstein held for his more elementary examples of mathematical practice, the apparent self-evidence of an example or figure is instead the result of training and conventions (BBH, 1996, pp. 182–183). That is, self-evidence must be replaced by social training if one is to understand how mathematicians agree about an object or inference. People do not arrive at the hospital already wearing perfectly fitting pyjamas, and the objects of Lakatos' narrative do not enter

[2] These studies formed the core of the strong program account of rule following. See Bloor (1992, 1997); Lynch (1992a, b). It is also possible to challenge these seemingly natural practices historically, as I have done in Barany (2014), which shows the arbitrary (and, indeed, deeply racist) development of ideas about the naturalness of certain approaches to counting and basic arithmetic in the second half of the nineteenth century.

the discussion obviously or inevitably as polyhedra or counterexamples to Euler's theorem. It is the hospital staff's or mathematicians' training and experience that helps them say (contingently, as each case arises) what is what. The processes of establishing mathematical definitions and procedures are thus viewed as coordination problems subject to social explanations. The mathematicians in Lakatos' tale always work with polyhedron-concepts derived from a finite stock of examples, principles, and intuitions, and so both "the counterexamples and the proof-idea had to be actively brought into contact with the concept of the polyhedron" (Bloor, 1976, p. 139).

By focusing on the dialogical elaboration of the concept of polyhedra and a theorem about it, Lakatos' story exemplifies what meaning finitist SSK takes to be the central process of mathematical knowledge-making. Here, polyhedra are understood in terms of a set of instances, what Barnes calls the concept's "tension" (Barnes, 1981, p. 308). The tension includes at different times many kinds of images of certain exemplary polyhedra, including those classified as cubes, prisms, and tetrahedra, as well as formal rules or definitions like "a solid whose surface consists of polygonal faces" (Lakatos, 1976, p. 14) and heuristics or principles such as "objects to which Cauchy's proof applies." There are necessarily only finitely many elements of this tension, and Lakatos' narrative can be read as a battle over what should properly be included. Every mathematical claim, here, is an assertion about what it means to be a polyhedron, and thus how the classification of "polyhedron" should apply. Conversely, when we talk about polyhedra (in general), we are implicitly talking about whatever composes the concept's tension, just as we speak of all the individuals wearing hospital pyjamas when we discuss patients in the abstract.

A new proposed definition or counterexample may be disqualified from deserving the title of polyhedron, though each one Lakatos introduces can plausibly be either accepted or rejected (meaning finitist claims 1 and 2). Figures that were once deemed polyhedra and proofs and definitions that were once thought to apply to polyhedra can lose that status in light of new examples or arguments (claim 3). In each case, new examples are evaluated in light of existing definitions and intuitions for what a polyhedron is (claim 4), and the examples are often also contrasted with or judged with respect to other related mathematical terms, like edges or surfaces (claim 5). When a rule or image is classified with the term "polyhedron," it is issued a pair of (always revocable) polyhedron pyjamas which enroll it in subsequent disputes over the status of polyhedra. Such a classification changes how these objects are viewed, understood, and used, temporarily emphasizing or attributing some of each object's features while obscuring others.

INDUCTION

Thus, in place of ideal objects to be described by rules and illustrated by examples, the meaning finitist account of mathematical objects gives us a heterogeneous constellation of images and formalisms. Conventional accounts of mathematics tend to take a top-down approach to the epistemology of mathematical objects, putting ideals at the centre and asking about the relationship between such ideals and what we know about them. By contrast, meaning finitism insists on a bottom-up approach, asking first how we know about objects and only later (if ever) inquiring after what those objects really are, if indeed they can be said to have an existence beyond what we know about them. Where a conventional account of Lakatos' story would say that different understandings and definitions of the same ideal polyhedron changed or were disputed, meaning finitists would find it unnecessary to stipulate the ideal polyhedron at all, and would instead make those changing understandings the centre of their version of the story. For the latter, the pyjamas make the polyhedron.[3]

Here, objects lose all connection to the ideal world and are instead examined solely in terms of how they are understood in the context of mathematical arguments. This is why Bloor asserts that "Polyhedra have no essence." Whether or not they have an essence in some ideal realm, his interest is in what we can know with certainty about polyhedra in the human world. Within meaning finitism, this means that the philosophical study of mathematical objects, which might include both ontological and epistemological concerns, becomes solely an epistemological inquiry into mathematical knowing to the exclusion of ontological questions about mathematical being. Even if we care foremost about what is underneath the pyjamas, the pyjamas are what we see, and hence what we study.

If all we care about are the acts of labelling associated with a concept, the concept itself seems to slip from view. Meaning finitists see it differently: if all we care about are the acts of labelling associated with a concept, then we may as well speak of concepts and their associated labels interchangeably. Because labels are always changing to incorporate new understandings and examples, this means that concepts, too, are always changing. The implications can be counterintuitive. Even though we may interact with people differently when they are issued (or not issued) different pyjamas, we do not usually think of those people as themselves changed by their clothes.

[3] I thank the editors for underscoring the analogy latent in the "pyjama" interpretation of polyhedra to the maxim that "the clothes make the man."

Meaning finitists ask us to see a person in one set of pyjamas as different from that "same" person in a different outfit because we interact with that person differently. People start to seem less like stable individuals and more like shape-shifting products of superficial interactions. The corresponding implications for the (in)stability of mathematical objects help explain the hostility towards the strong program from many philosophers of mathematics, whose discipline was in many respects founded on the distinction between objects and their representations.

On the one hand, it is not necessarily strange to think of mathematical concepts as existing in a constant state of flux. In Lakatos' tale, processes such as "monster barring" allow mathematicians to exclude examples from consideration, and "monster adjustment" allows them to domesticate instances to fit certain formal specifications by recasting their salient features (Lakatos, 1976, pp. 14–23, 30–33). As Lakatos makes clear, something that counts as a polyhedron for some people at one point in time has no guarantee of counting as one for different people or at different times. Disagreements about classifications and properties of specific postulated examples show that even if there were a "right answer" that existed outside of mathematical negotiations, it would not be decisive for the historical elaboration of a theorem like the one Lakatos describes.

But while what counts as a polyhedron is not fixed for Lakatos, the formalisms of mathematics appear relatively more so. The process of proofs and refutations, for Lakatos, is a dialectic whose end is formalisms which are better able to circumscribe the shadowy apparitions of the informal polyhedron-concept. Lakatos thereby introduces an asymmetry, between informal and formal objects or concepts, that appears at odds with the tenets of the strong program of SSK, which assert that *all* objects comport to the same social rules of classification. A more detailed elaboration of how mathematical objects are developed through mathematical practice is needed to account for this apparent asymmetry under the uniform rubric of meaning finitism.

All knowledge, for meaning finitists, has the same basic starting point: "Induction is constitutive of human thought at every level" (Barnes, 1974, p. 9; see also Bloor, 1976, p. 118; Barnes, 1981, p. 320). Because "there are no terms for which meaning or use is self-evident," even mathematical terms must be developed from something that is, of necessity, non-terminological (Barnes, 1982, pp. 26–27). No matter how complex, every concept can be traced to some primary acts of ostention, where initial terms are baptized in relation to elements of one's immediate experience. That is, words, objects, and experiences are neither self-generalising nor self-defining, and must be understood in terms of other words, objects, and experiences. One's experience with the body of phenomena we call the physical world undergirds a "highly elaborated world-picture" containing meanings, practices, and conventions "only tenuously connected with what can fall within anybody's experience" (Bloor, 1976, p. 86; Bloor, 1983, p. 91; Bloor, 1997, p. 39). Our concept of circles, for instance, comes about only through repeated encounters with definitions, illustrations, properties, and proofs (see BBH, 1996, pp. 63–64). As Lakatos illustrates for polyhedra, this process of elaboration can in principle be open to contestation at every step, and, indeed, it is often that very contestation that drives the elaboration forward.

Patterns in our experiences allow us to form generalisations. In this view, Euler's theorem that $V-E+F=2$ generalises the expectation that whenever we encounter a polyhedron, it can be found to satisfy the relation $V-E+F=2$. (Note the active "can be found to" in place of a passive "will" – having a mathematical property is always the result of an active determination.) This is a clear example of epistemic induction, anticipating that the future will conform to the patterns of the past. In mathematics, as in most things, such an inductive proposition is generally tacit. Induction concerns assessments of probability and confidence in generalisations (Barnes, 1981, pp. 318–319). Mathematics, the art of making particularly confident assertions about particular kinds of patterns, is thus an extreme example of the inductive thinking at the heart of all concept formation.

Of course, new information and new understandings can alter the meaning of a theorem like the one Lakatos considers. Despite its changing meaning, we can still understand it as the same theorem from one moment to the next. This continuity is maintained in two basic ways. First, one can often match new objects to the pattern "polyhedron" with little difficulty, and can confirm (or challenge) the inductive theorem on the basis of that identification. Though the theorem may have been modelled with a picture of a tetrahedron, a picture of a cube or triangular prism will not trouble an ordinary attempt at verification. But there remain cases where this pattern-matching is less clear. In this second situation, the theorem is modified or upheld by establishing a convention regarding the new object (or alternatively by modifying the theorem), as when Lakatos' characters debate whether a new proposed counterexample is really a polyhedron.

Concepts, like the theorems about them, follow the same bipartite system of maintenance. Thus, Lakatos' characters most often count vertices, faces, and edges routinely – there is little dispute about simple tetrahedra or other familiar shapes. Quite frequently, however, objects are much harder to recognise unequivocally. Confronted, for instance, with polyhedra that have stars rather than convex polygons as faces, Lakatos' characters undergo a process of monster adjustment, arguing over different ways of identifying what an edge or a face really is with respect to a specific example. These two systems of theorem- or concept-maintenance represent two kinds of classification activity in strong program meaning finitism, drawing primarily on either observable pattern matching based on identifiable features of objects (natural-type) or social negotiation based on conventions of identification and signification (social-type).[4] Some people – perhaps those wrapped in bandages – just look like hospital patients, even before you issue them pyjamas. Some people require examination and fitting, and may require that the pyjamas be altered in some way, before the hospital clothes will be both a social and sartorial fit.

The asymmetry in Lakatos' treatment of formalisms derives in part from this distinction. Where assessments of mathematical statements appear to take the form of a natural-type classification, the flexibility and conventionality of these statements' interpretations is elided, in part, by the success of the pattern-recognition. Through much of Lakatos' book, his characters work to establish ultimately arbitrary conventions in order to clarify concepts that are initially less formal. These appear to readers as social-type classifications because it is easy to identify the presence of ambiguities and competing interpretations. On the other hand, activities like counting and identifying edges and vertices are particularly well practiced and rarely controversial (even if used to innovative or unusual ends), so formal propositions involving these terms have a strongly natural-type appearance, even where they may have been initially bootstrapped and maintained by social-type activities.

Formalisms thus achieve their apparent stability through the naturalization of social-type classification into a counterfeit of natural-type classification. For instance, that the results of an algebraic deduction appear inevitable to trained mathematicians is seen to be a natural property of algebraic formalisms rather than the result of a learned system of conventions and manipulations that could have been otherwise. The case of star-polyhedra and the technique of monster adjustment from Lakatos' narrative, however, shows how even firmly stabilized patterns can be tested and reconfigured (e.g. Lakatos, 1976, pp. 16–17, 30–33).

[4] Barnes calls these, respectively, P/N devices and S/S devices (Barnes, 1983, p. 530). Bloor (1997, p. 40) goes further to distinguish pattern-matching activity from its mathematical interpretation, in which case all that is properly mathematical is maintained through S/S devices. As with the view painted here, Bloor nonetheless makes sure to emphasize that mathematical knowledge depends inescapably on the interpretation of *both* natural and social kinds, and hence on both P/N and S/S devices.

Here, examples and models have an especially important place. They are the basis of both learning old concepts and developing new ones (Barnes, 1982, pp. 18, 52; BBH, 1996, pp. 102–103, 105; Bloor, 1997, p. 11). Because pattern matching and analogy in abstract mathematics are based, to only a limited extent, on perceptual judgements, there are more opportunities for intervention from social-type stipulations of similarity or dissimilarity (BBH, 1996, pp. 106–107). Whether an object gets to wear pyjamas in mathematics depends more on what we decide to think of it than on how it may look to us. Where perceptual judgements are employed in mathematical arguments, they are almost invariably assessments of model images or forms meant explicitly to stand in for an abstract infinity of ideal objects: we make judgements about ideal abstract triangles based on the images we can draw on a piece of paper or a blackboard. Simple, manageable examples take a particular prominence because of their greater workability and presence in our empirical experience (Bloor, 1976, p. 90).

Indeed, the premise in meaning finitist epistemology that knowledge is empirical and inductive puts a heavy premium on the immediate, familiar, and workable. In mathematics, objects are said to exist on a full scale of conceptual accessibility from the most basic ostensible representations, to less basic but still operable and workable formalisms and examples, all the way to the infinities of utterly ungraspable ideal objects. Strong program treatments of mathematics focus on precisely the simplest and most immediate cases because, in addition to being more accessible for sociologists and philosophers, they are presumed from the start to be at the heart of the greater problematics of mathematical knowledge. If the simplest and most obvious concepts are fundamentally social in their constitution, the reasoning goes, then so must be the more complex ones.

As Lakatos illustrates, the social negotiations at play in our knowledge of 2+2 reappear at nearly every juncture in the attempt to characterize polyhedra. If 2+2 is irreducibly social (and meaning finitists would point to the considerable room for variation in the meaning and use of numbers and the operations of arithmetic to insist that it is), then the compounded conventional character of polyhedra should be doubly ineliminable. For 2+2 is a simple formalism tied to simple ostensive experience, whereas "polyhedron" is a highly mediated concept encompassing a range of heuristics, formalisms, and heuristics and formalisms about those heuristics and formalisms. Because the "polyhedron" concept comes from a much broader tension, it must fit each instance that much more slackly, and be that much more elastic. At the same time, meaning finitists risk a certain amount of question-begging. Put crudely, it is not surprising that an approach beginning with the central importance of simple examples should conclude that those same examples are centrally important in the constitution of (mathematical) knowledge. From a meaning finitist perspective, this does not present a problem, for it is taken as a matter of principle that all concepts have the same sorts of epistemic foundations and are formally indistinguishable.

But in a more conventional view, there is a substantial difference between the knowledge of 2+2 and that of V−E+F. The former is a formalism understood to stand in for an infinity of possible empirical instantiations. The latter is also a formalism, but at its root it is a formalism about formalisms. It describes a putative relation among an infinity of polyhedra, themselves comprehended as mixtures of empirical images and formalisms. Moreover, all but a few of the ideal polyhedra to which Euler's theorem is taken to apply are known from the start to be utterly unrealizable. At the root of the problem of induction is the question of what can count as future instances of a phenomenon.

PLATONISM

As mathematical claims and concepts grow more complex, their significance emanates more and more from mathematicians' ability to reason with and extend them beyond their initial contexts of enunciation. Euler's theorem is not just about images of cubes, though such images are prominent at the beginning of Lakatos' narrative. Rather, Euler's theorem is about a vast constellation of polyhedra, both realized and unrealized. Just as it is possible for a new theorem to contradict "already known" examples, a mathematician may also use a theorem to make statements about "as-yet unknown" examples with some justifiable confidence. It is even possible for theorems to imply meaningful assertions about objects which are not strictly knowable, such as polyhedra too complex to characterize explicitly. Lakatos shows the problems for mathematical objects at their conceptual boundaries, but it can be argued that every single formalism in his account nonetheless applies completely unproblematically to an infinite collection of ideal polyhedra in what might be called the concept's conceptual interior. Hospitals stock the range of pyjamas that they do because those pyjamas will fit most patients without difficulty.

This view is a form of mathematical Platonism, a matter of central concern in SSK (e.g. Bloor, 1983, p. 83). Platonism is both an ontological and an epistemological position. Ontologically, it holds that mathematical objects exist independent of human interventions. There will always be ideal pyramids, for example, and it will always be the case that V−E+F=2 for them, regardless of where the Ancient Egyptians buried their royalty or whether Euler performed any calculations or made any conjectures. Epistemologically, it holds that we can obtain (albeit always imperfect) knowledge of such ideal mathematical objects through reasoned deductions. In both respects, Platonism overlaps substantially with realism, another of SSK's pivotal touchpoints, which holds that objects in nature exist independent of our experience of them (e.g. Bloor, 1973, p. 176; BBH, 1996, p. 88, *et passim*).

Unsurprisingly, meaning finitists are fundamentally skeptical of Platonism. Orthodox Platonism implies that the collection of objects to which a mathematical rule or statement applies is fixed in advance and that concepts have stable essences, contradicting the flexibility tenets of finitism (Barnes, 1982, p. 32; Bloor, 1983, pp. 28–29, 88; BBH, 1996, p. 85; Bloor, 1997, pp. 37, 130). With its implication that some statements are pre-given as correct, Platonism undermines SSK's principle of symmetry – that agreement about theories should be explained with the same social mechanisms, whether those theories are deemed true or false (Bloor, 1973, pp. 176–177; Bloor, 1997, p. 36).

But the meaning finitist objection to Platonism comes with a twist. While there is no hope of reconciling meaning finitism and Platonism on epistemological grounds, the distinction between Platonist ontology and Platonist epistemology allows scholars of SSK to dispense with the latter without altogether doing away with the former. Platonist epistemology, they argue, is irredeemably circular. Even if there are essential truths about ideal objects in mathematics, one cannot know that one has found them except through social mechanisms such as argumentation and demonstration (Bloor, 1973, p. 182; Bloor, 1983, p. 86). Meaning finitists uniformly reject Platonist epistemology, just as they reject its realist cousin.

When it comes to ontology, however, Platonism becomes just one of many possible outlooks, implying its own special set of strategies for managing knowledge (Bloor, 1997, p. 38). The sociologist of science need not judge the ultimate reality of the objects of knowledge, but should rather study the social function of positing certain things as real (Bloor, 1973, p. 190; Barnes, 1982, p. 82). We behave differently towards pyjama-wearing entities in the hospital because we believe them to be humans with life histories and everything else that entails, and (as long as we believe that) it would not much change our behaviour if those entities were *really* mechanical automata or particularly convincing sacks of rags underneath. That is not to deny their humanity, but to say that when it comes to our actions it is our socially defined belief in their humanity that really counts, rather than any purported underlying reality of humanity. In the case of polyhedra, it is not necessary to say whether or not there are such things as ideal polyhedra or eternally valid theorems about them that are independent of human activity. Rather, one starts from the perspective that mathematicians are human, and that as humans they make a variety of claims (Platonist and realist ones) that shape the mathematical objects and theorems under discussion. One does not say whether or not ontological Platonism is valid while at the same time recognizing that such Platonism is a widely held view that fundamentally shapes the kinds of knowledge claims mathematicians can and do make.

Indeed, meaning finitists seem perfectly happy to endorse a "naive common-sense realism" positing an independent reality but, in keeping with the ontology-epistemology distinction, "refusing to conflate external reality with anything that is said of it" (BBH, 1996, p. 88). For Barnes (1982, p. 79), realism with respect to nature is directly analogous to Platonism in mathematics, and the latter is even used as a justification of the former. On evaluating a large number of otherwise unproblematic polyhedra and finding that for several of the more complicated ones $V-E+F$ came out to three, one would suppose oneself to have miscounted or to have inadequately represented the object rather than reject the original proposition. As Barnes suggests, just because some eggs would fall out differently on different runs through an egg-sorting machine it is not necessary to reject the proposition that eggs come in different sizes (Barnes, 1983, p. 541). Strong relativist claims about *how* we know remain, in this sense, impartial with respect to *what* we know.

In this sense, it is still consistent with meaning finitism to claim, as is famously attributed to mathematician Henri Poincaré, that geometry is "the art of reasoning correctly about figures which are poorly constructed" (see Netz, 1999, pp. 33–34; Hardy, 1967 [1940], p. 125). One has only to admit the instantiating role of figures without claiming for them unmediated access to any essence in what they depict. In terms of their process of elaboration, mathematical objects always go beyond their Barnesian tension of instances, not by having an ultimate (ontological) essence but by being (epistemically) flexible, yet relational, in their future applications. A putative ideal shape with millions of vertices, edges, and faces can, as a practical matter, only count as a Platonic polyhedron to the extent that it is possible to compare it with the simple rules, figures, and heuristics in the polyhedron-concept's tension.

That is, there is no immediate way to verify one way or another whether such a shape *really* exists, but meaning finitism helps us account for the epistemic process by which mathematicians can confidently believe that it does, as well as for the effects that belief has on their claims and practices.

Polyhedra, whether they exist Platonically or not, maintain a social existence compatible with Platonism by virtue of their meaning finitist constitution. Because mathematicians understand polyhedra through piecemeal experiences, intuitions, models, and heuristics, it is always possible to suppose that there is some unified ideal concept underneath it all. Polyhedra are pyjamas all the way down, but that is precisely what makes it possible to believe in the reality of that which the pyjamas classify. As with our account of induction in meaning finitist mathematics, we again find simple workable examples at the core of concept-formation.

CONCLUSION

In the end, Barnes' pyjama analogy reminds us that labels must be made to fit their objects. Hospital pyjamas are loose-fitting but not omni-fitting. Similarly, in Lakatos' story, the "polyhedron" label has an inherent flexibility that allows genuine differences of understanding over what can count as a polyhedron without implying that any object whatsoever could readily be so labelled. Indeed, the inductive character of meaning finitist mathematical knowledge suggests that the negotiations that must occur at the contested boundaries of the polyhedron-concept can only apply to simple, workable candidate polyhedra. We can understand V−E+F to equal 2 for infinities of unpicturable polyhedra, but in the cases where the polyhedron-concept really counts – those cases at the heart of mathematical research and understanding – the exemplars in question must be representable in ways that make the label accessible and meaningful.

In an important practical sense, most ideal polyhedra on the interior of the polyhedron-concept cannot wear pyjamas – there is no lived situation in which they can be directly manifested, manipulated, addressed, and labelled. Such wholly putative polyhedra subsist unproblematically precisely insofar as they are never called into question. Indeed, most of them cannot, as a matter of practice, be put to the test. The meaning finitist account of Platonism helps us recognise that most polyhedra exist in a meaningful sense only because they can be posited in relation to workable guiding instances. These latter instances correspond to what Lakatos seems to indicate with the term "heuristics." We tend to think of heuristics as guideposts to understanding what mathematics is really about, but for a meaning finitist all we can really know and explain are the guideposts themselves – those objects that can be issued pyjamas at all.

Nor can hospital pyjamas be made to stretch and pull without bearing traces of such contortions. Mathematicians generate new concepts and refine old ones by balancing the familiar with the unfamiliar, and each decision about the scope and meaning of an idea affects what is possible for future ones. Like pyjamas, labels can tear at the seams. Old mathematical frameworks or definitions can prove untenable in the face of contradictory intuitions or heuristics. Concepts can fail to win approval under the weight of countervailing arguments – this is Lakatos' process of monster barring. In each case, classical SSK meaning finitism affords an enriched view of the resources and processes that make conceptual adaptation in mathematics possible.

As an alternative to epistemic Platonism, SSK meaning finitism forces renewed attention onto the role simple examples play in even the most complex mathematical knowledge. This focus on simple objects and their labels, born as much out of methodological necessity as principled conviction, yields under further consideration a robust framework for interrogating the mathematical uses and meanings of models and heuristics. It helps one take account of what is manifestly evident in mathematics, both past and present: that the discipline's objects and notions are shaped, challenged, and manifested through the social interactions of mathematicians.

ACKNOWLEDGMENTS

This material is based in part on work supported under a National Science Foundation Graduate Research Fellowship (Grant No. DGE-0646086) and under a Marshall Scholarship. I thank Pablo Schyfter and Jane Calvert for their helpful early comments on this essay, and the editors of this volume for their more recent insightful suggestions.

REFERENCES

Barany, M. J. (2014). Savage Numbers and the Evolution of Civilization in Victorian Prehistory. *British Journal for the History of Science*, 47(2), 239–255.

Barnes, B. (1974). *Scientific knowledge and sociological theory*. London: Routledge.

(1981). On the Conventional Character of Knowledge and Cognition. *Philosophy of the Social Sciences*, 11(3), 303–333.

(1982). *T. S. Kuhn and social science*. London: Macmillan.

(1983). Social Life as Bootstrapped Induction. *Sociology*, 17(4), 524–545.

Barnes, B., Bloor, D. & Henry, J. (1996). *Scientific knowledge: A sociological analysis*. London: Athlone.

Bloor, D. (1973). Wittgenstein and Mannheim on the Sociology of Mathematics. *Studies in History and Philosophy of Science*, 4(2), 173–191.

Bloor, D. (1976). *Knowledge and social imagery*. London: Routledge.
(1978). Polyhedra and the Abominations of Leviticus. *British Journal for the History of Science*, 11(3), 245–272.
(1983). *Wittgenstein: A social theory of knowledge*. London: Macmillan.
(1992). Left and Right Wittgensteinians. In A. Pickering (Ed.), *Science as practice and culture* (pp. 266–282). Chicago: University of Chicago Press.
(1997). Wittgenstein, rules and institutions. London: Routledge.
Hacking, I. (2014). Why is there philosophy of mathematics at all? Cambridge: Cambridge University Press.
Hardy, G. H. (1967 [1940]). *A mathematician's apology*. Cambridge: Cambridge University Press.
Lakatos, I. (1976). *Proofs and refutations: The logic of mathematical discovery*. Cambridge: Cambridge University Press.
Lynch, M. (1992a). Extending Wittgenstein: The Pivotal Move from Epistemology to the Sociology of Science. In A. Pickering (Ed.), *Science as practice and culture* (pp. 215–265). Chicago: University of Chicago Press.
(1992b). From the 'Will to Theory' to the Discursive Collage: A Reply to Bloor's 'Left and Right Wittgensteinians'. In A. Pickering (Ed.), *Science as practice and culture* (pp. 283–300). Chicago: University of Chicago Press.
Netz, R. (1999). *The shaping of deduction in Greek mathematics: A study in cognitive history*. Cambridge: Cambridge University Press.

2

Mathematical Concepts? The View from Ancient History

REVIEL NETZ

INTRODUCTION

To clarify the scope of this chapter, I start with a quotation, which the rest of the article will criticize. Not that I have any special gripe against this quotation! To the contrary: I take it precisely because it is from a competent and well-received study in mathematics education. I wish to convey some skepticism concerning a mode of argument in the history of mathematics, organised around "concepts".

So, a quotation from a study that aimed to replicate the historical conceptual evolution of "actual infinity" among schoolchildren:

> Infinity emerged as a philosophical category in the work of Aristotle, but not yet as a "mathematical object". The potential character of infinity is found in the Aristotelian conception. The actual infinite character of the natural number sequence is not considered. We shall later see that before this could become possible, the concept of set had to be incorporated into mathematics (Luis et al., 1991, p. 212).

Here is a view according to which the Greeks did not study actual infinity because they did not have the concepts required for discussing it.

Let me say something in general about "concepts". The term is new (Greek philosophy does not have any word to translate it easily) and is fundamentally an artifact of twentieth-century philosophy. There, two parallel observations were made, by philosophers of mind on the one hand and by logicians and philosophers of language on the other. The philosophers of mind point out Brentano's thesis of intentionality: "Every mental phenomenon includes something as object within itself." Thoughts are mental, and there ought to be some mental object for thinking about, say, numbers. We are not just automata that respond to stimuli (such as utterances) in correct ways. Hence, we need to have, say, "the concept of number", or else, just what would your thinking be constituted from? The logicians, separately, point out Frege's famous distinction between "meaning" and "reference". *The Morning Star* and *The Evening Star* both refer to the same lump of matter – the planet Venus – but their identity of matter does not make them into an identical object; they must differ, therefore, as concepts. Hence, our logic requires the notion of concepts, associated (but not in a one-to-one relationship) with references.

One sense of "concept", then, is that connected with the workings of the mind. To put this crudely, it exists in an individual's head. The other is that associated with the functions of logic and language, whereby it exists in an abstract sense.

Historians, who refer to the evolution of concepts, engage with a somewhat distinct category. What is intended by a "concept", in an example such as that cited above concerning the concept of infinity, is some kind of widely shared understanding of a certain object, that acts as a condition of what can be said concerning it. Let us bear this in mind: the strong arguments for the existence of "concepts" (those of Brentano, and of Frege) have little to do, directly, with the kind of object required by the historiography of "concepts".

I have started out with an example concerning infinite number and indeed it is inside the history of number that the model of the "concept" is most often applied by historians of mathematics. Not only is it a widespread, nearly universal category, but it is also a very flexible one. This flexibility is very familiar – roughly, that of an expanding balloon.[1] At first, with the Greeks, the balloon catches very little: only the positive integers (excluding *one*). Then, the balloon begins to expand: fractions are added, then one and zero, then negative numbers; the continuum is filled with irrational numbers, and an entire new dimension of complex numbers is added (further dimensions may be added as we expand "number" into more complicated algebraic structures); the balloon, finally, expands to take in infinite numbers as well. We start with a minimal concept, end with a maximal one. This is a valid, if rough, outline of the history of mathematics (with a few qualifications that follow).

In the explanatory model based on "concepts", the domain of Greek "number" is seen to be restricted. It referred only to concrete countable collections – that many coins, that many sacks of grain. In particular, the Greeks did not have a concept that ranged simultaneously across countable collections as well as continuous magnitudes (no concept ranging

[1] This is now best covered in Corry (2015).

across both integers and real numbers), and for this reason they were forced to approach geometry and arithmetic as entirely separate mathematical fields. It was only later, with the rise of a more powerful concept of number, that the unification of number and geometry was made possible. This Cartesian unification was the first significant abstraction in the history of mathematics, enabling our own modern, general and abstract mathematics (one in which, indeed, even infinity is a number). The history of the concept of number is then the decisive element in the history of mathematics.

Such is the outline of a standard story, standardly set out in Klein (1934/1936/1968). In Netz (2002), I argued against it. I titled my article then "It's not that they couldn't", arguing against the historiographical strategy of conceptual impossibility, where it is claimed that, "They did not do X, because they did not yet have the concept". This strategy, I claimed, fails empirically. What the argument from conceptual impossibility predicts is the absolute absence of a certain practice. In fact, what we tend to find historically is not that certain practices are absolutely impossible at different times, but rather that they are comparatively *rare*. It's not that they couldn't: it's just that, more often than not, they didn't.

In this chapter, I will explain my doubts concerning this explanatory model. I will first talk about infinite number, then recapitulate the argument of Netz (2002), and then offer a tentative account concerning concepts in history.

INFINITY, ACTUALLY

Luis and colleagues essentially repeat a commonplace (that I was taught, and believed, myself, before I was forced otherwise by the evidence). Let me quote again:

> Infinity emerged as a philosophical category in the work of Aristotle, but not yet as a "mathematical object". The potential character of infinity is found in the Aristotelian conception. The actual infinite character of the natural number sequence is not considered. We shall later see that before this could become possible, the concept of set had to be incorporated into mathematics.

First of all, this is a valid description of Aristotle's philosophy. He produces an account of change, space and time, where there is no need for infinity and infinitesimal ever becoming actual, and instead all one requires is the potentially infinite (so that the line may always be bisected, a number may always be added, etc.[2]).

Let me add more background. There is a Greek mathematical technique, attested first in Euclid's *Elements* Book XII (probably due to Eudoxus, one of the most influential early Greek mathematicians), and then used very widely by Archimedes, referred to as "The Method of Exhaustion". Suppose, for instance, one wishes to calculate the volume of a cylinder. Then, have a prism, erected from a polygon at one of the bases, inscribed inside the cylinder. Clearly the volume of the prism would be less than that of the cylinder. By making the polygon have more sides, we can also make D – the difference in volume between the cylinder and the prism – smaller and smaller. We need to prove that for any given magnitude M, we can make the difference D smaller than that magnitude M. This calls for geometrical ingenuity, but once we achieve a proof of this kind (for any M, a smaller D may be constructed), we can provide an argument – similar to Cauchy's delta–epsilon proofs – deriving the volume of the cylinder. (Cauchy, indeed, essentially builds upon the ancient Method of Exhaustion which, in this way, is simply part of the modern toolbox of the calculus.) So, all in all, the above citation is valid: there is a central Greek philosophical argument, against the existence of actual infinity; a central Greek mathematical technique, eschewing this infinity.

Aristotle, Eudoxus, Archimedes: important names. And yet, the most significant part of the citation above is its title, "Infinity in Greek Culture": the statement about Aristotle is *directly* taken to be a claim about the Greeks as a whole. This is not so absurd if one's starting point is the study of conceptual evolution: first, you had this concept, then you had that one … so that it is assumed that there should be "the Greek concept of infinity" preceding our own – and if so, we might as well take a single well-known Greek specimen to deduce the concept of Greek culture as a whole.

In fact, the historical picture is more complicated, even for the most canonical authors mentioned above. Already Kouremenos in 1995 and even more directly Knorr in 1996 have pointed out that, in fact, besides the very celebrated Method of Exhaustion, there are also more marginal passages in Greek mathematical writings where the actual infinite seems to be entertained. This was made even clearer, in dramatic fashion, in 2001–2002, thanks to readings from the Archimedes Palimpsest.

[2] There is a difficulty here, in that typically things that exist potentially in Aristotle do so because they sometimes exist also in actuality. Lear (1979) offers an argument that we should indeed understand "potentially" in this case strictly, so that Aristotle does deny the existence of actual infinite sets or divisions.

I was lucky to be involved in the reading of this – the central extant source for the works of Archimedes. The Archimedes Palimpsest was originally made in about 975, probably in Constantinople and originally was a collection of several treatises by Archimedes. In the year 1229 it was recycled, its leaves taken apart and scraped, with the text of a prayer-book overwriting the text of Archimedes. It was rediscovered in 1906 by the great Danish philologer J.L. Heiberg, who managed, remarkably, to read most of the underwriting. The manuscript was then stolen and resurfaced yet again in 1998. Among the first new readings made was that of Archimedes' *Method* 14, and in what follows I sum up the results of Netz, Saito and Tchernetska (2001, 2002).

Already Heiberg was able to reconstruct most of the proposition. Archimedes measured the volume of a cylindrical hoof. We look, to begin with, at Figure 2.1: the base of a cylinder inscribed within an upright prism (whose top base is not shown in the figure).

The cylindrical hoof is produced by (a) producing the cylinder on the circle HZEΘ) and (b) extending the inclined plane through the diameter HE and to the parallel edge of the top of the prism, above the line ΓΔ. The solid intercepted between the semicircle HZE, the inclined plane, and the cylinder, is the cylindrical hoof (see Figure 2.2). We also produce, inside the same base, the parabola HΛZE (which is uniquely determined by the three points H, Z and E).

We now cut the solids through an arbitrary plane, orthogonal to the base, whose footprint on the base is as the straight line NM. The line NΛ is cut off by the parabola (parabola line); the line NΣ is cut off by the circle

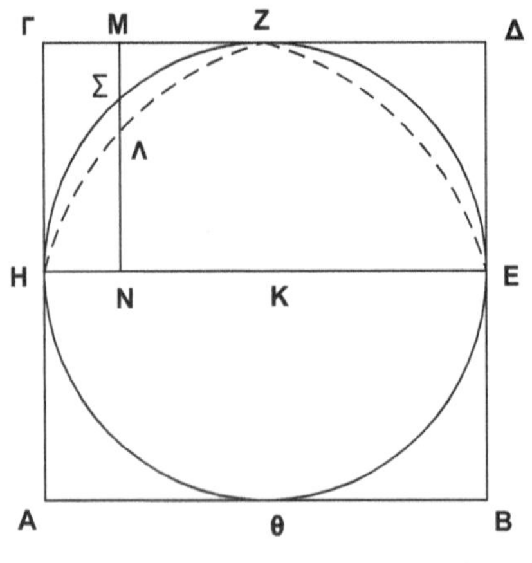

FIGURE 2.1 The Base of a Cylinder Inscribed within an Upright Prism.

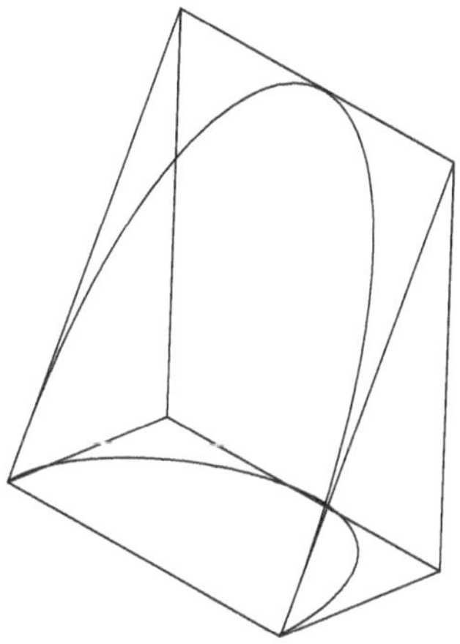

FIGURE 2.2 The Cylindrical Hoof.

(circle line); the line NM is cut off by the prism itself or the rectangle at the base (rectangle line).

Archimedes finds first the following marvelous geometrical relation:

$$(\text{rectangle line}) : (\text{parabola line}) :: (\text{rectangle line})^2 : (\text{parabola line})^2$$

This is stated by Archimedes without proof, and so I leave it to you to prove, as well, and I provide a clue in a footnote.[3] It directly follows from this that the triangle above line NM is to the triangle above line NΣ as the rectangle line is to the parabola line.

So far, Heiberg could read; but at this point the tiny, erased letters no longer made sense, and he picked up the thread of the argument only more than half a column later, where it was stated that the triangles such as those above the line NM constitute a triangular prism, those above the lines NΓ constitute the cylindrical hoof, the rectangle lines constitute the rectangle and the parabola lines constitute the parabolic area, so that all in all we have a proportion with four terms:

triangular prism : cylindrical hoof :: rectangle : parabolic area

And since the parabolic area has been measured in a previous proposition, we can therefore derive the volume of the cylindrical hoof as well.

This then seems like a suggestive heuristic argument where one passes directly from the result in planes to a result in solids, which already definitely

[3] Try to look for right-angled triangles within which you can apply Pythagoras' theorem. Also, remember the way in which the Greeks primarily conceived of a parabola: as an arena for equalities between squares and rectangles.

suggests a willingness to engage with actual infinity (and was cited as such by Knorr, 1996).

In 2001, we could read the actual argument Archimedes produced to support this derivation. It reads as follows:

> There are certain magnitudes equal to each other – the triangles in the prism; and there are other magnitudes, which are lines in the parallelogram[4] ΔH, being parallel to KZ, which are both equal to each other, and equal in multitude to the triangles in the prism; and those triangles, in the segment cut off, shall also be equal in multitude to the triangles that come about in the prism, and the lines taken away from the lines drawn parallel to KZ between the section of the right-angled cone[5] and EH, shall be equal in multitude to the <lines> drawn parallel to KZ in the parallelogram ΔH.

I underline the words "equal in multitude", which shall interest us especially in what follows.

While Archimedes does not make this explicit (he is not very explicit in general: remember the lack of proof for the main geometrical result), it is clear that he applies here a general lemma, developed elsewhere (as the first proposition to another treatise, *On Conoids and Spheroids*), that allows him to sum up many ratios. The lemma shows that when certain conditions hold between all the terms of the sets A, B, C and D ("sets" is not Archimedes' own word), then the sum of the terms in set A to the sum of set C is as the sum of B to the sum of D. The sets in question, in our case, are the triangles in the prism and in the cylinder and the lines in the rectangle and the parabola.

The proof in *On Conoids and Spheroids* 1 assumes that the four sets are all finite. Here, however, they are infinite: there are infinitely many slices of the prism, hence infinitely many triangles.

How many precisely? Archimedes notes that these are, in fact, *equal* to each other *in multitude*:

> lines in the parallelogram to triangles in the prism;
> triangles in the cylinder to triangles in the prism;
> lines in the parabola to lines in the parallelogram.

The background to this expression is as follows. In Greek geometrical texts, the subject of discussion is often lengths, areas and volumes and it is then stated, for instance, that "line AB is equal to CD", "triangle ABC is equal to

[4] Archimedes refers to the rectangle as a "parallelogram".
[5] "Section of the right-angled cone" is Archimedes' expression for "parabola".

square DEGH", "three times the cone is equal to the cylinder". Usually, such expressions of "equality" are used without qualification. Sometimes, there is a need to distinguish this sense of equality from equality in cardinality. Thus, in the citation above, it is asserted that "the triangles in the prism are equal to each other", the intended meaning (in our terms) being that they all have an equal area. This is distinguished from the statement that the same triangles are equal "in multitude" to the triangles in the cylinder. (In other contexts, Greek mathematicians may call the first kind of equality "equality in magnitude", to distinguish it from "equality in multitude"). Since the expression "equal in multitude" is formulaic (though everywhere else referring to a finite sense), we have no reason to read "in multitude" to mean here anything different from "in number".[6]

To sum up: Archimedes considers three distinct pairs of infinite sets and asserts that the two infinite sets in each pair are equal to each other in multitude, which in context means precisely the same as "equal in number".

What does this mean? Now, it is clear that Archimedes in general develops propositions such as *Method* 14 as a heuristic exercise and that he does not consider the argument as a whole to be rigorous. But there is no indication that the logical flaw resides in the very claim of equality in number between the sets; instead, this would most naturally reside in the application of this claim, to transition from the infinitely many individual proportions involving planes to the one summed up proposition involving solids. The application of *On Conoids and Spheroids* in this context is indeed problematic, and Archimedes could have been aware of this. But it is perfectly possible that he considered the statement of equality in multitude between the triangles as innocuous – as indeed it is. We will note immediately that a one-to-one relationship can be geometrically produced for the triangles (every triangle in the prism contains uniquely a triangle in the cylinder), and while the phrase "one-to-one relationship" is anachronistic, the geometrical intuition of equality would be accessible even without this piece of modern set-theory terminology. And at any rate, even if Archimedes entertained doubts about the very claim of equality, the text before us

[6] There are simple stylistic reasons to prefer "in multitude" to "in number". (a) Greek uses here the simple dative form to express a manner: the sets are "equal" *plēthei*, "in multitude" (Dat.), and the equivalent expression "equal" *arithmōi*, "in number" (Dat.) would most naturally be read not as "equal in the manner of number-hood" but simply as "equal to a number". (b) *plēthei* is contrasted, in this formulaic expression, with *megēthei*, "in magnitude". *Megēthos* and *plēthos* form a more natural pair of two words used in highly abstract contexts (they even *sound* the same, which probably matters here), while the natural range of the word *arithmos*, "number", is more specific and concrete.

certainly puts forward a proposition in which doubts may be entertained. Archimedes, quite directly, discusses the number of an actually infinite set.

Now, I am not saying that the concept of infinite numbers was widespread in antiquity. To the contrary: it was extremely rare. I do not think it is an accident that the clearest example of an actual infinity used by an ancient mathematician was read only in 2001–2002, from a unique palimpsest. It had to wait for so long precisely because it was so rare (which made it, indeed, so hard for Heiberg even to read: he generally could read well, where he could rely on patterns established elsewhere). Infinity was generally avoided in Greek mathematics, for a rather straightforward reason: in Greek mathematical texts, rigor is highly valued, and it is hard to develop the mathematics of infinity without using, at least at first, some heuristic shortcuts. But it's not that they couldn't. When Greek mathematicians wished to consider actual infinity, they did so. Hence, what was at stake was not some kind of an absolute conceptual barrier, but simply a matter of preference.

The statement that "The actual infinite character of the natural number sequence is not considered [in Greek mathematics …] before this could become possible, the concept of set had to be incorporated into mathematics" is therefore, I think, simply wrong. It is not that the Greeks lacked the concept of "set" and therefore could not develop the modern concept of "actual infinity" and therefore could not discuss this. It is simply that they usually preferred to discuss other things. And the point, conversely, is not that the Greeks in fact did "possess the concept of actual infinity" because they did "possess the concept of set". Rather, what we see is that, as far as the historical account is concerned, the very category of "concept" seems immaterial. The discussion of an actual infinity is conceptually easily attained by Archimedes, and there is no barrier there. What stops him from doing more infinite-talk is not the lack of a conceptual structure, but the preference for rigor over an expressive mathematical language. And indeed the thing is precisely that language and definitions (conceptual structures that is), do not matter to Archimedes. He cares about mathematical arguments and ultimately values, above all, proof. When this is reversed among some authors of the seventeenth century, more experimental uses of infinity begin to sprout, leading on to the calculus. It is not that at some point a new concept was acquired; rather, a certain path became more attractive, as mathematical values changed. This, I argue, is not an accident shown by this single example: rather, this is the norm in the history of number. While definitions of number come and go, they do not seem to constrain the kind of mathematics produced in different eras.

INDEED, IT'S NOT THAT THEY COULDN'T

At this point let me quickly recapitulate the examples produced in Netz (2002), all having to do with the concept of number:

1. Fractions. Do the Greeks have a general "concept of fractions"? To do so – argued Fowler (1987, pp. 264–5) – we will need to see an example of manipulating a fraction (say, *n*th of *m*) *as a fraction*,[7] for instance by stating that "the *n*th of *m* multiplied by the *q*th of *p* gives *nq*th of *mp*". Fowler argued that there was no such example, and produced this as evidence for the lack of a concept. Christianidis (2004) pointed out that there *was* such an example, in fact – though he could point out only one: Diophantus, Arithmetic IV.36. This is, of course, exactly analogous to the example developed above for infinity: not the absolute absence predicted by an account based on concepts, but simply something which is *very rare*. (So why do Greeks use unit fractions so often, general fractions so rarely? Because unit fractions were the standard form: a convention of shared practice, not the constraint of a given concept.)
2. Arithmetized geometrical objects. A key "conceptual" innovation, if you will, of Cartesian geometry is the arithmetic manipulation of geometrical terms – for instance, their multiplication (so that two lines multiplied "are" the rectangle). This is not the practice of Greek geometry where rectangles are referred to as geometrical objects, not as quantitative multiplications.[8] I pointed out in Netz (2002) that this important generalisation does have some exceptions, in particular in an expression used by Archimedes towards the end of *On the Sphere and Cylinder II*, which can most naturally be read to mean "the square multiplied by the line" (to derive a solid volume). This, however, appears in a specialized, marked text, perhaps to emphasize its special characteristics. (Once again, then, we should understand the

[7] Otherwise, we could take expressions such as "*n*th of *m*", say "a third of two" as referring simply to *m* times the unit fraction "*n*th" or "two times third" (that the Greeks, as well as many other early civilizations, did possess such unit fractions is uncontroversial, and the historical debate concerns the rise of the general concept of a fraction, as an expansion of an earlier concept of numbers as just integers, from which unit-fractions may be derived. The main discussions are Knorr (1982), as well as the contributions by Fowler and Vitrac in Benoit, Chemla and Ritter (1991)).

[8] This is the key claim of Unguru (1975) (who therefore argued for a strict conceptual separation between ancient and modern mathematics), a claim that was controversial when first made and is now universally accepted.

strict geometrical reading of geometrical objects as a convention of practice, not as a conceptual barrier.)

3. Quantified geometrical objects. Not only are Greek geometrical objects not manipulated arithmetically; quite simply, they are not treated, in most elite Greek mathematical texts, as possessing numerical value at all. Thus, an author such as Euclid does not make his square have, say, the length of "two feet". While this is true for most elite Greek mathematics, however, the same is no longer true in the texts of Hero of Alexandria, where lines, areas and volumes are given numerical values. In Netz (2002), I saw this as a more general case of the one above: I argued that just as Archimedes wished to mark a certain passage in *On the Sphere and Cylinder II* apart by the use of arithmetical manipulations, so Hero as a whole wished to mark himself apart as an author. I was somewhat wrong. What I failed to realize then was the tight connection between Hero's corpus and the Greek practice of mathematical *education*, where quantified geometrical objects were apparently routine.[9] The simplest account of Hero's corpus, I now believe, is as a teacher of the exact sciences, seeking the elite status associated with book authorship. Notice an important consequence of such a scenario for our understanding of the relationship between concept and mathematical training. It is not that the Greeks did not possess a unified concept of number that could range across discrete and continuous entities, hence they could not do an arithmetical geometry. To the contrary: they were even *trained* in a practice where numbers were routinely assigned to geometrical entities and, only then, those of them who turned to produce geometry of the elite sort learned the convention of eschewing number in the context of elite geometry.
4. Is one a number? One of the features often noted about the "strange" nature of the Greek concept of number is the exclusion of one, the unit, from "number" (which is perhaps not entirely surprising if indeed, as I suggest, "number" and "multitude", in the Greek context, are effectively synonymous). Indeed, Euclid's definitions of "one" and "number" are unambiguous. But it is a mistake to take definitions at face value. In Netz (2002), I simply cited the important contribution eventually published as Høyrup (2004), who has pointed out the very

[9] This is based on an understanding of the Greek mathematical papyri as belonging strictly to education (as opposed to being remnants of an ancient mathematical literature). In this, I follow Sidoli (2015): compare Netz (2011).

simple fact that Euclid himself defines "perfect numbers" as those equal to the sum of their parts, and then of course includes 1 among those parts (otherwise, 6 would not be a perfect number!). Many readers respond by suggesting that 1 could be a part of 6 without being a number (but then why is a half not a part of 6?), but clearly this is besides the point. Euclid has different concerns in mind, when he defines "number" and when he defines "perfect number"; the potential tension between the two definitions is not among his concerns. An important lesson: do people – even mathematicians, even in their axiomatic moments – even care about what their concepts "really are"?

5. An infinitely extendible series of numbers. My concern here is with the Greek system of numerical notation which was evidently bounded, unlike our own open-ended decimal system. (The central Greek system of numerical notation has 27 symbols stand for the numbers 1 ... 9, 10 ... 90, 100 ... 900, and then use a special mark to turn those into thousands; the system is theoretically bounded just beneath one million.) I pointed out in Netz (2002) that the Greek abacus was, in fact, an indefinitely extendible system of numerical representation, exactly analogous, structurally, to a decimal positional system of writing. That the Greeks did not transfer this system from the operational abacus, to the written text is important: but it is clearly a matter not of conceptual impossibility, but rather of the choice of media: practice, not concepts.
6. Finally, I pointed out in Netz (2002) that there is an overarching narrative as if Greek mathematical texts tend to eschew the numerical, as such, concentrating instead on qualitative geometrical relations, so that entire fields of mathematics (such as combinatorics) are absent. I then pointed to the important study by Acerbi (2003) that has shown that there must have been ancient texts in combinatorics; Netz et al. (2003) has since added a potential example of a geometrical calculatory text by Archimedes, and in Netz (2009) I have looked in more general into the field of calculation in Greek mathematics. In this particular case, the general impression is simply misleading, or perhaps a construct of a particular form of reception, that tends to emphasize certain aspects of Hellenistic mathematics more than others.

In short, it's not that they couldn't. When we look closely at Greek mathematics, we do not see a homogeneous set of texts, all bounded by the conceptual poverty of their "number". We see a range as diverse as that of any mathematics up to and including the seventeenth century, distinguished from more modern mathematics simply by its preferences and emphases, above all its practices. Concepts, once again, are simply beside the point.

WHAT WE SHARE

I began this essay by conceding that "concepts" are required both for the philosophy of mind as well as for the philosophy of language, because it is difficult to explain mental and logical phenomena otherwise. I have then argued, essentially, that concepts are beside the point for the history of mathematics. This seems like an obvious tension, and it appears strange that we should have concepts as mental and logical categories, and then toss them out as we move on to science.

While counterintuitive, this is not quite absurd. The argument from mental phenomena, as well as from logic, requiring that there should be mental concepts, does not yet require that these should be historical agents. To take first the argument for concepts as logical entities, this essentially involves language or a structurally similar system of conceptual structure – and the kind of argument that demands that such a system would be a historical agent would be essentially along the lines of the Sapir-Whorf thesis, of language constraining thought. There is no need to rehearse here the well-known weaknesses of this thesis in its strong form.[10]

As for concepts as mental categories, their existence is derived strictly speaking from a phenomenological argument, set within a single mind.[11] Indeed, it seems perfectly plausible that different individuals will have different mental props sustaining their reasoning on the very same topic. The phenomenological argument for the existence of mental concepts does not involve a shared category at all.

Notice, however, that as we move to larger logical structures such as propositions and proofs, intersubjectivity emerges. A proposition has the property that it is either true or false, a proof has the property that it is either valid or invalid, and such truth values – while potentially contested – are understood (under most realist assumptions) to be external

[10] The best measured statement of the state of this problem – sympathetic to the restricted sense in which the Sapir-Whorf thesis might contain a grain of truth – is Deutscher (2010).

[11] The empirical evidence for concepts as a meaningful psychological category is, in fact, mixed. It is not that this is not a useful term: rather, it seems to be very heterogeneous, manifested in a host of unrelated ways in the mind (for instance, sometimes we use prototypes, sometimes we rely on classifications): on the basis of this diversity, Machery (2009) advises that philosophy, itself, should be "Doing without concepts".

to the mind. It is therefore natural that the category of the "proposition" or of the "proof" should be primarily intersubjective. Thus, there has to be a widely shared understanding of what is meant by "the number of all prime numbers is greater than any assigned number", as well as a widely shared understanding of what constitutes a proof of such a claim, even without there being a widely shared understanding of what is meant by "number".

The idea that a molecule may have a property that none of its constituent atoms possesses is strange; but obviously there are such "emergent" properties. This becomes even more evident when we consider relative terms, such as "being shared intersubjectively". So, to take an obvious example, in the transition from the microscopic to the macroscopic levels, objects gain the emergent property of *visibility*. This is more or less a direct analogue to the way in which the property of intersubjective sharing emerges in the transition from concepts to propositions. One's individual concept is, in a sense, invisible (present only to a single mind, not to a collective with shared understandings). Of course, we may have discourses about terms and their meanings, and in this way make them visible (intersubjective) – but what becomes visible (intersubjective), then, is not so much the underlying concepts themselves, but rather the entire discourses concerning them, distilled into the – propositional – structure of a definition.

If indeed concepts, in the mental sense, are not shared intersubjectively, it makes perfect sense that they should not be explanatory in science. Science is what people do together, not what they each have privately, in their heads. Indeed, the main line of historical change involves, quite simply, the different ways in which things are done together – the transitions of scientific practice, which have nothing to do directly with concepts. The trend to greater abstraction and generality in modern mathematics is real enough. It derives from a combination of historical changes (such as the new intersections between mathematics and philosophy, and then the academicisation of mathematics) that have nothing to do with conceptual change: and as mathematicians came to engage with different practices, and came to produce different theories with different kinds of proofs, so, in varying ways, did their concepts change.

What to make of this argument for mathematical education? If my argument presented in this chapter has any merit in it, what transpires is that this question is difficult to answer, for a precise reason. Mathematical education engages with the interface between the private and the intersubjective – between that which individuals have in their head and that which societies achieve. For the first, concepts are crucial; for the latter, they are almost an afterthought. History suggests the sheer difficulty, if not impossibility, of the task of instilling a concept in a pupil's head, because we use intersubjective tools – to obtain private, and murky, conceptual growth.

And if we still wish to distill a positive lesson from all of this, after all, I would suggest the following. The importance of understanding is evident. I do not recommend the blind drilling of procedures. But this much can be said: if the history of mathematics is any guide for the classroom, it seems to suggest that what matters is not so much the understanding of mathematical concepts, but rather *the understanding of mathematical methods and practices.*

REFERENCES

Acerbi, F. (2003). On the shoulders of Hipparchus: A reappraisal of Ancient Greek combinatorics. *Archive for History of Exact Sciences*, 57(6), 465–502.

Benoit, P., Chemla, K. & Ritter, J. (1992). *Histoire de fractions, fractions d'histoire.* Boston, MA: Birkhäuser Verlag.

Christianidis, J. (2004). Did the Greeks have the notion of common fraction and did they use it? In J. Christianidis (Ed.), *Classics in the history of Greek mathematics* (pp. 331–336). Boston, MA: Springer.

Corry, L. (2015). *A brief history of numbers*. Oxford: Oxford University Press.

Deutscher, G. (2010). *Through the language glass: Why the world looks different in other languages.* New York: Metropolitan Books.

Fowler, D. (1987). *The mathematics of Plato's Academy: A new reconstruction.* Oxford: Clarendon Press.

(1992). Logistic and fractions in early Greek mathematics: A new interpretation. In Benoit, P., Chemla, K. & Ritter, J. (Eds), *Histoire des fractions, fractions d'histoire* (pp. 133–147). Boston, MA: Birkhäuser Verlag.

Høyrup, J. (2004). Conceptual divergence – canons and taboos – and critique: Reflections on explanatory categories. *Historia Mathematica*, 31(2), 129–147.

Klein, J. (1934/1936/1968). *Greek mathematical thought and the origin of algebra.* Boston, MA: MIT Press.

Knorr, W. (1982). Techniques of fractions in ancient Egypt and Greece. *Historia Mathematica*, 9(2), 133–171.

(1996). The method of indivisibles in ancient geometry. In R. Calinger (Ed.), *Vita mathematica: Historical research and integration with teaching* (pp. 67–86). Washington, DC: The Mathematical Association of America.

Kouremenos, T. (1995). *Aristotle on mathematical infinity.* Stuttgart, Germany: Franz Steiner Verlag.

Lear, J. (1979). Aristotelian infinity. *Proceedings of the Aristotelian Society,* 80, 187–210.

Luis, E., Moreno, A. & Waldegg, G. (1991). The conceptual evolution of actual mathematical infinity. *Educational Studies in Mathematics*, 22(3), 211–231.

Machery, E. (2009). *Doing without concepts.* New York: Oxford University Press.

Netz, R. (2002). It's not that they couldn't. *Revue d'Histoire des Mathématiques*, 8(2), 263–289.

(2009). *Ludic proof: Greek mathematics and the Alexandrian Aesthetic*. Cambridge: Cambridge University Press.

(2011). The bibliosphere of ancient science (outside of Alexandria). *NTM Zeitschrift für Geschichte der Wissenschaften, Technik und Medizin*, 19(3), 239–269.

Netz, R., Acerbi, F. & Wilson, N. (2004). Towards a reconstruction of Archimedes' *stomachion*. *SCIAMVS*, 5, 67–99.

Netz, R., Saito, K. & Tchernetska, N. (2001). A new reading of Method Proposition 14: Preliminary evidence from the Archimedes Palimpsest (Part 1). *SCIAMVS*, 2, 9–29.

Netz, R., Saito, K. & Tchernetska, N. (2002). A new reading of Method Proposition 14: Preliminary evidence from the Archimedes Palimpsest (Part 2). *SCIAMVS*, 3, 109–129.

Sidoli, N. (2015). Mathematics education. In W. Bloomer (Ed.), *A companion to ancient education* (pp. 387–400). Oxford: Wiley-Blackwell.

Unguru, S. (1975). On the need to rewrite the history of Greek mathematics. *Archive for History of Exact Sciences*, 15(1), 67–114.

Vitrac, B. (1991). Logistique et fractions dans le monde hellénistique. In Benoit, P., Chemla, K. & Ritter, J. (Eds) (1992). *Histoire des fractions, fractions d'histoire* (pp. 149–173). Boston, MA: Birkhäuser Verlag.

PART II

3

Notes on the Syntax and Semantics Distinction, or Three Moments in the Life of the Mathematical Drawing

JULIETTE KENNEDY

INTRODUCTION

The mathematician's life revolves around proving theorems, mainly, but at the same time the mathematician is a creator of visual objects, of *drawings*. These mathematical drawings – what to make of them? Artifacts of the main action – the proof – they have, usually, no lasting relevance. In certain cases, though, the idea of the proof cannot be understood without the drawing, either in the form of a concrete token or in the form of a mental image. The proof, in such cases, seems to be somehow inherently pictorial.

In this chapter I attempt to frame a theory of the mathematical drawing. I suggest that, seen from what we will call the "single text" point of view, such drawings help us towards a different understanding of mathematical practice, one that assumes direct access to mathematical content; one that resists the bifurcation of that content into its (possible) semantic and syntactic modes.

Two Subgenres

The first thing to say is that the genre of the mathematical drawing seems to admit two categories. There are the Euclidean diagrams – *metonyms for epiphany*, or so they have been called,[1] created by the movement of the hand, but directly constitutive of the proof (see Figure 3.1).

And there are the informal, illustrative drawings, as one might call them, the *incidental* drawings mathematicians create while working alone or in conversation with other mathematicians, or while giving lectures (see Figure 3.2).

Depending on what part of the scholarship one consults, the Euclidean diagram oscillates between being a proof and being a set of instructions for

[1] Catton and Montelle (2012, p. 27). See discussion that follows.

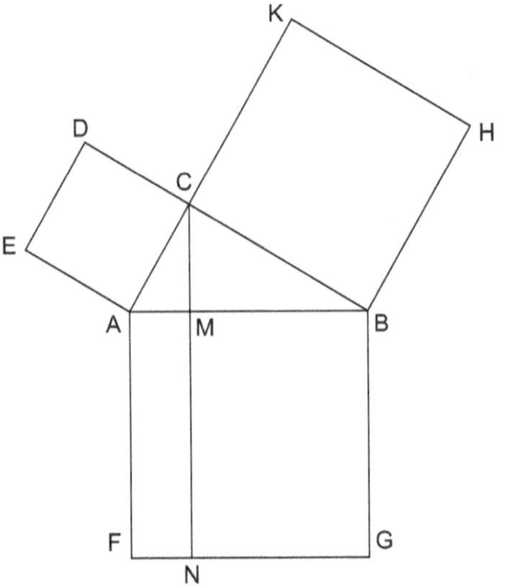

FIGURE 3.1 A diagram showing the initial construction for the proof of Book I, Proposition 47 of Euclid's Elements (the so-called Pythagorean theorem).

FIGURE 3.2 Drawing by Andreas Villaveces and Timothy Gendron.

a proof – a bidirectional capability that is the source of extraordinary philosophical opportunities, as K. Manders has said (1995/2008, p. 68)– and the diagram is accompanied by a text, a set of equations given in a sequence, directives to the reader to observe *congruences*.

How does this mixture of language and image produce knowledge, for us, and for the Greeks? The question is complicated by this peculiar ambiguity that surrounds Greek diagrammatic practice, the question whether the drawing is a proof or not. Even the Greek word *diagramma* has two meanings: diagram in the usual sense of an instructional image, and "proposition":

> This is a notorious fact about Greek practice: it is generally difficult to tell whether the authors speak about drawing a figure or about proving an assertion, and this is because the same words are used for both. And this again is because the diagram is the proof, it is the essence of the proof for the Greek, the metonym of the proof.[2](Netz, 1998, p. 38)

There is a substantial literature on the Euclidean diagram. What I call *incidental* diagrams, on the other hand, have not been the object of much critical attention. This lack of attention is odd, considering the fact that the production of visual imagery is an important, one might even say essential, part of mathematical practice.

What can be said about these incidental diagrams? Considered in the context of the discourse to which they are associated, they do not encode its logic so much as compress that discourse into an image, a visual *mathematical.*[3] Unlike the Euclidean diagram, which is radiant with absoluteness, the incidental diagram reminds us of all that is raw and unfinished, *unrigorous* and incomplete in mathematics. Taken as single moment in the production of a piece of mathematical knowledge, they give us the mathematician in the grip of insight *à l'état brut*; they are the trail of the human serpent in the garden.[4]

This chapter contains a sample of such drawings, Euclidean drawings and others, along with, here, a consideration of three moments in the life of the mathematical drawing: *Euclidean, architectural* and *incidental*, or respectively: the drawing as proof object, but heaven-sent, as it were, a token of the realm of pure ideas; the drawing in the light of the

[2] "Pappus, when commenting on earlier mathematical treatises, estimates the size of a treatise by the number of diagrams. He does not say only 'this work has 73 propositions': rather, he often says 'that work has 73 diagrams'. Of course, this shows first of all something about the Greek word *diagramma* which Pappus uses. It means much more than just "a diagram." It means something much closer to "a proposition" or "a proof"" (Netz, 1998, p. 37). See also Netz (1999).

[3] The noun "mathematical" is due to A. Arana, and is meant to be "inclusive of objects, concepts, structures, whatever we talk about when we talk about, think about, mathematics" (personal communication).

[4] William James (1978): "The trail of the human serpent is [...] over everything" (p. 37).

manual-practical agency of the architect; and finally the drawing as the carrier of lived sensibility.

FIRST MOMENT: THE EUCLIDEAN DIAGRAM

Consider the case of the classical Greek mathematical diagram. From Catton and Montelle (2012, p. 27):

> [T]he Greek mathematical diagram is a metonym for an epiphany, namely, an epiphany concerning manual practice, by which one rationally resolves how to prosecute the relevant agency, so that in consequence new concepts are formed.

The classical Euclidean proof, under this view, involves a text, but it also involves the action of the hand under leadership by the text, an action in which, by a kind of "manual-practical leading," the student draws their way towards a "needed practical epiphany" – draws their way, that is, towards an understanding of the proof:

> The neophyte learns only by engaging with manual actions under leadership by the text ... in order truly to learn from Euclid, one needs one's hands, not only one's eyes and one's brain. This is at least to say that either manually or in the (practical) imagination, one must move about a diagram as one constructs it – yet also it must move one too, to a timeless insight. Otherwise one is not managing to do geometry with the Greeks. (p. 27)

Catton and Montelle (2012) build on K. Manders' (1995/2000) groundbreaking work, an analysis of the role of the diagram in Euclidean proofs that calls into question earlier, critical conceptions of Euclid's diagrammatic practice.[5] Manders' question – how can it be that the Euclidean drawing can serve as a carrier of inferential weight? – is answered there in terms of what he calls the diagram's *exact* and *co-exact* properties, corresponding, roughly, to the diagram's incidental and invariant properties (respectively). When a diagram displays a triangle, for example, it displays a co-exact property; if a line segment occurring in the diagram has a particular magnitude, then having that magnitude is an exact property of the diagram.

[5] It has long been known that in Book I of Euclid's *Elements*, the proofs require more than what is stated in the Common Notions and Postulates. A statement of the view Manders nevertheless wishes to challenge is the following: "Euclid did not have an axiom system in mind, and did not develop geometry axiomatically, in Book I of the *Elements*." See Seidenberg (1975).

Manders observes that, inferentially speaking, co-exact properties always trump exactness. Thus, the Euclidean diagram encodes a kind of pictorial logic, and Manders' observation is that it is only by virtue of its co-exact properties that the diagram contributes to the proof.[6] As Manders puts it, "Euclidean demonstration has no alternative justificational resources" (p. 62).

In addition to Catton and Montelle (2012), Manders (1995/2008) has either given rise to or coincided with other developments. Shin (1994) has argued in her work that the Euclidean diagram draws on a visual syntax supporting a special diagrammatic logic which can be studied in just the way that any other logic can be studied, i.e. for soundness, completeness and so forth. And others have formalised this logic in different ways, either *reconstructively*, as in Avigad, Dean and Mumma (2009), or in ways that do not prioritise deference to the original texts, i.e. seeking merely to extract a diagrammatic logic which is both correct and workable.[7]

This belief in the inferential power of the Euclidean diagram is new – or again new. Descartes instigated a revolution in mathematics by treating geometric concepts algebraically, and although this was an essential step in the development and rigorization of mathematics, this would suppress the Euclidean diagram, not only pedagogically but *ontologically* – that is, in its very function within the proof. The judgement became entrenched in the practice subsequently, so that in 1882, Moritz Pasch would describe his axiomatisation of geometry as effecting the banishment not only of the "figure" but of sense altogether:

> In fact, if geometry is genuinely deductive, *the process of deducing must be in all respects independent of the sense of the geometrical concepts*, just as it must be independent of figures; only the relations set out between the geometrical concepts used in the propositions (respectively definitions) concerned ought to be taken into account.[8] (quoted in Schlimm, 2010, p. 98; emphasis added)

Pasch's view that true rigor does not allow for a drawing was adopted by David Hilbert in his 1894 work, raising the question whether the

[6] This turns on the fact that the co-exact properties of a diagram are shared by all instances of it. Thus, a single diagram can stand in for all configurations that lie in the range of the proof. For more about the so-called generality problem, see Shin (1994). The question how a single diagram can support a general result, how a proof of a proposition relating to a specific diagram can be thought of as generalising to all possible diagrams of the same kind, was considered already by Aristotle.

[7] For an overview of the recent literature on diagrammatic reasoning in geometry and other areas, see Shin, Oliver and Mumma (2014).

[8] See also Shin, Oliver and Mumma (2014).

banishment of "sense" and the subsequent rupture between syntax and semantics in the twentieth century – a division of labor, so to speak, to which the modern logician is nowadays completely bound – began with the suppression of drawings. I return to this point later in the chapter.[9]

Pasch's remark needs to be understood against the background of an expanding algebraic discourse in mathematics – but this is just one way to move against the diagram. It is well known that Plato abjured thinking of geometry in practical terms, as a matter of "doing things";[10] as if the activity of the geometer were centered on the production of impermanent and temporal images – for what is more ephemeral than a drawing? – than with the timeless and acausal mathematicals of which we have true knowledge.

It is interesting that Pasch's view did not suppress diagrammatic practice outside of geometry. New diagrammatic systems were introduced at the time by logicians, most notably Frege and Peirce, who are credited (independently) with introducing, in the form of these systems, modern quantificational logic. Analytic philosophers have been mostly concerned with first-order logic as it is conceived apart from any diagrammatic presentation of it – until recently, when Manders' analysis of the Euclidean diagram stoked interest in *logical* diagrammatic practice. Thus Frege's logical diagrams have been considered by Danielle Macbeth (2014), who asks in that work (in analogy to the question about Euclidean diagrams): How do Frege's logical diagrams serve as vehicles for the acquisition of *logical* knowledge? The answer, according to Macbeth, has to do with those features of Frege's logical practice which involve reasoning *in* the diagram and not *about* the diagram, a "distinction between describing or reporting a chain of reasoning in some natural language and *displaying or embodying* a chain of reasoning" (p. 73; emphasis added). As for Peirce's logical practice, many have read into Peirce's diagrammatic logic, and the "flow of experience" which supports it,

[9] For an interesting later counterpoint to the "suppression of drawings point of view" see Hilbert and Cohn-Vossen's (1932/1952) *Geometry and the Imagination*. See also Arana's (2016) discussion of the work.

[10] Arana (2016) claims, Plato's views "were at odds with Greek geometric practice" (p. 465). Nevertheless, Plato's influence was sufficient to drive into the discourse surrounding the diagram, and more widely into the discourse surrounding mathematical practice altogether, a conception of mathematical concepts as unembodied, timeless and acausal, and mathematics itself as descriptive, a set of truths about an objectual domain. As Arana describes Plato's view, "if geometry is to produce knowledge rather than mere opinion it must shun apprehension by construction with its apparent temporality" (p. 465).

a similar, ontological reliance on the diagram – a *logical* practice in which the diagram is, embodies or displays the proof.[11]

I am concerned here with the mathematical drawing, Euclidean and incidental drawings, and the question, how can it be that the act of drawing an image, either manually or in the practical imagination,[12] can deliver mathematical understanding, when this should be a matter of pure analysis?[13] If the diagram's *rational moment*, the "needed epiphany," in Catton and Montelle's (2012) terminology, is the outcome of reason in its "manual-practical" mode, how is this rational moment accomplished? How is it that pure analysis, on its own, cannot get us there?

Catton and Montelle's radical answer is this: Insofar as reason is restricted to its analytic mode, or, for Catton and Montelle, its *logical* mode, reason must be in a special sense *undone*; and space must be cleared for the operation of reason in its synthetic mode:

> Whether the diagram's rational moment chiefly concerns inference – that is to say, whether the diagram chiefly assists rational progression from some already articulate thoughts to some other – we question … If we seek to understand the nature of the reasoning in Greek geometry, it is good neither to equate reasoning with logic, nor even to equate it with inference of a diagram-assisted kind from some articulate thoughts to some other. To understand Greek geometry in its own intellectual form … is to consider the synthetic function of reason – as deeper-lying and still more important within Greek geometry than the analytic function of reason (p. 27).

Their wish to sever reason from analysis, or even, perhaps, from logic, creates difficulties everywhere. But their question whether a notion of inference can be developed that harmonises with (what they have called) manual/practical agency – the primary capability, they argue, that enables the construction of the Euclidean diagram – is an

11 It is beyond the scope of this chapter to treat the diagrammatic practice either of Peirce, whose diagrammatic practice has sparked a large literature, or of Frege.

12 This is the distinction in what Giaquinto (2015) calls "visual thinking," i.e. "thinking with external visual representations (e.g., diagrams, symbol arrays, kinematic computer images) and thinking with internal visual imagery; often the two are used in combination, as when we are required to visually imagine a certain spatial transformation of an object represented by a diagram on paper or on screen."

13 For the purposes of this chapter here and subsequently I understand the pure analytic functioning of reason in a very strict sense, as purely syntactic and void of content. An alternative view would not see the pure analytic functioning of reason as void of content in this sense.

interesting one. Of course, traditional theories of inference have always asked whether (and how) reason can function not only purely analytically but in a way that makes essential use of other, synthetic, or synthetic-constructive modes of reasoning. The line of thought goes back to Proclus, whose commentaries on Euclid were widely read, also in the modern era, who invoked the *imagination* in order to explain constructive modes of reasoning in geometry – the imagination in the two-dimensional sense that is, thought of as a faculty which presents or represents the diagram to the mind, generating projections "as if on a screen" so that the understanding might act upon them.[14]

The idea that the imagination enables synthetic reasoning wove itself into the veil of philosophy from then on. It would be greatly elaborated by Kant, who devised a complex theory of synthetic modes of judgement in mathematics, built around the analytic-synthetic distinction, and the further parsing of these two categories of judgement into the *a priori* and the *a posteriori*.[15] Kant's move inaugurated the *semantic tradition*, so-called, in which synthesis came eventually to be thought of as having to do with *meanings*; and the synthetic functioning of reason came to be thought of as a faculty of intuition, in which such meanings could be grasped.[16]

The subsequent, *post*-Kantian development is what concerns us here, in particular Bolzano's separation of contentual or *world-involving inference* from mere logical consequence in his monumental 1837 work *Wissenschaftslehre* – a distinction that has survived and more than that is utterly central to logic nowadays, the distinction between semantic and syntactic consequence.

The syntax/semantic distinction does not come without critique, Quine's (1951) paper (questioning the analytic-synthetic distinction altogether) being perhaps the most well known. Of interest here is the idea that such a critique might be motivated by considerations of diagram use in mathematics. This is the idea that the diagram-assisted proof functions as a *single text*, an *ensemble intégré*; and thus the rational moment enabled by the diagram is *prima facie* indecomposable in Bolzano's sense; i.e. the inference cannot be split into its syntactic and its semantic modes.

14 The comparison of the imagination with a screen is due to Proclus. See Arana (2016).

15 Kant defined synthesis as "the act of putting different representations together, and grasping what is manifold in them in one cognition" (A77/B103); a process that "gathers the elements for cognition, and unites them to form a certain content" (A78/B103).

16 Our view of the emergence of the semantic tradition is in agreement with that of Coffa (1991) in his landmark work *The Semantic Tradition from Kant to Carnap: To the Vienna Station*.

I take a moment to explain what is meant by this idea. The following example appears in Avigad et al. (2009):

> Let L be a line. Let a and b be points on L, and let c be between a and b. Let d be between a and c, and let e be between c and b. Is d necessarily between a and e? (p. 705)

As the authors remark:

> [I]t is hard to make sense of the question without drawing a diagram or picturing the situation in your mind's eye; but doing so should easily convince you that the answer is "yes." With the diagram in place, there is nothing more that needs to be said. *The inference is immediate, whether or not we are able to cite the axioms governing the betweenness predicate that would be used to justify the assertion in an axiomatic proof system* (p. 705; emphasis added).

If we let *P* be the proposition of the above proof, that *d* is between *a* and *e*, then what Avigad and colleagues are noting here is that the inference to *P*, the principal (and indeed only) inference of the diagram, is delivered as a matter of pure visual perception.

This is a simple example. But if one considers the various forms of inference at play in more complex Euclidean diagrams, then it is possible that the principal inference associated with these diagrams consists of an aggregate of inferences of the form, inference to *P*. The diagram-assisted proof is, in my terminology, enabled by a collection of *single texts.*

Are other propositions in mathematics assimilated as single texts? Possible candidates might include the pigeonhole principle, which states that if n items are placed in k containers, with k<n, then at least one container must have more than one item. Another example might include the well-ordering of sets of natural numbers.

When does the aggregate – text plus visual mathematical – fail to cohere as a single text? A theorem of Bolzano (1817) states that if a continuous function assumes a negative value at a real number *a*, and a positive value at a real number *b*, then it must have a zero in the interval $[a, b]$. The statement of the theorem follows immediately from the relevant diagram – and yet it does not.[17] Establishing the theorem requires the least upper bound principle (in fact the two statements are equivalent). Thus, the relevant inference

[17] See, for example, Courant and Robbins: "Here for the first time it was recognised that many apparently obvious statements concerning continuous functions can and must be proved if they are to be used in full generality." (1941, p. 312).

is not enabled by a single act of perception – *despite every appearance that it does.*

Resolving the difficulty requires a nuanced view of Bolzano's motivation for proving the theorem in the first place: the search for explanation in the form of objective grounds, along with a conception of the mathematical discipline of analysis, that seeks to ground its theorems strictly outside of geometry. As Bolzano wrote:

> There is certainly no question concerning the *correctness*, nor indeed the *obviousness*, of this geometrical proposition. But it is clear that it is an intolerable offence against *correct method* to derive truths of *pure*, or general mathematics (i.e. arithmetic, algebra and analysis) from considerations that belong to a merely *applied*, or special part, namely *geometry* … For in fact, if one considers that the proofs of the science should not merely be *confirmations*, but rather *justifications*, i.e. presentations of the objective ground for the truth concerned, then it is self-evident that the strictly scientific proof, or the objective reason, of a truth which holds equally for *all* quantities, whether in space or not, cannot possibly lie in a truth which holds merely for quantities that are in *space*. (in Russ, 1980, p. 160)

Bolzano goes on to say that, construed as a geometric truth, the theorem consists of a number of component concepts, or "simple truths … reasons for other truths and never themselves consequences." A theorem, then, has its basis in these simple truths, and thus "must be proved by a derivation from these other truths."

To summarise Bolzano's view – one that is now entrenched in the mathematical community – construed as a geometrical truth, the theorem follows immediately from the relevant diagram. Construed as a general mathematical truth, the theorem requires proof. The single text point of view is not new. Foucault (1966/2009) reads the language/world binary prior to the sixteenth century as no binary at all, but as a single text. As Curtis Franks (2014) has written in the opening of his "Logical completeness, form and content: an archaeology", quoting Foucault:

> The signifying function of words in the sixteenth century, according to Foucault, depends not on our acquaintance with them, or with their use, "but with the very language of things …" [A]ll the thinkers of the Renaissance … were "meticulously contemplating a nature which was, from top to bottom, written," for they inhabited "an unbroken tissue of words and signs, of accounts and characters, of discourse and forms." (pp. 59, 36, 40)

In the seventeenth century, we are told, things were different. Language was believed to be arbitrary, its relationship to the world contingent on the details of its fallible design and conventional use. "As a result," Foucault urges, "the entire episteme of Western culture found its fundamental arrangements modified. And, in particular, the empirical domain which sixteenth century man saw as a complex of kinships, resemblances, and affinities, and in which language and things were endlessly interwoven – this whole vast field was to take on a new configuration." (1966/2009, p. 54)

> We are instead in doubt about what before could not meaningfully be questioned: whether our systems of signs adequately fit the world. (p. 78)

Of interest here – my starting point – was Manders' work on diagrammatic reasoning, and on the heels of his work Macbeth's extension of it to logical diagrams. What is missing? From Manders:

> The conception of the role of diagrams in traditional geometrical thought given here should be understood as incomplete, probably on philosophically central points. While I go beyond inferential reconstruction in traditional logical terms' the roles to which we in the twentieth century are sensitive and know to articulate are broadly but constrainedly 'inferential'. It is plain that we are missing a lot. (1995/2008, pp. 81–82)

The mind/body binary that Plato drove so deeply into epistemology, the idea of the body as a presence that corrupts and confuses, and in every other possible way works against rational thought – an idea which nowadays expresses itself in the thought that the body is, at best, epistemologically irrelevant – is still with us. Philosophical discourse on the imagination draws heavily on visual metaphors, as we saw, the idea of the imagination is a faculty enabling a special kind of two-dimensional representation, not in the sense of an embodied capability but representation in the sense of projection onto a screen.[18]

But we now know that the synthetic, or active imagination – if we still want to call it that – is more complex. Thus it seems plausible that the imagination is in some way essentially embodied, and that the production of

[18] Though the active imagination appears already in Proclus, in that "the understanding draws upon the screen, and that is an activity: an activity of mind, but an activity ... modeled on the physical activity being represented ... older views of imagination [did not take] it for a merely receptive facility." Arana, personal communication.

internal imagery exceeds the two dimensional, being experienced, again, bodily.

de Freitas and Sinclair (2012) construe thinking altogether, as bodily experience *outright*:

> According to phenomenological currents within this [empiricist JK] tradition, thinking and reasoning, and any other related cognitive constructs, are always external or located in the flesh; "Thinking is not a process that takes place 'behind' or 'underneath' bodily activity, but is the bodily activity itself. (Nemirovsky & Ferrara, 2009). (p. 135)

And they too concentrate on manual agency:

> In studying a student's tactile and multi-modal engagement with a cube, Roth [(2010)] shows how the movement of the hands erupts or emerges without intention or governing concept. The haptic encounters are somehow more originary than language, somehow detached or free from the "knowing" that is bound to signification. It is in the hand that the memory of prior encounters with cubes is immanent. Roth suggests that there is a more originary pre-verbal "I can" that coordinates this encounter with the cube, and that the world begins to emerge through touch and the coordination of movements of eyes and hands. He privileges the movement of the hand itself, its "auto-affection", as an embodied activity that is prior to all verbal framing. (p. 136; emphasis in the original)

This leads us to the second moment in the life of the mathematical drawing: the drawing in the light of the manual-practical agency of the architect (see Figure 3.3).

SECOND MOMENT: "WHAT IS THE HAND?"

In his monograph *The Thinking Hand* – an ode to manual agency – the architect Juhani Pallasmaa (2009) gives us, indeed celebrates, the hand. "[A] prodigious precision instrument that seems to have its own understanding, will and desires," (p. 25) more than any other part of the body, it is the hand that serves as a bodily interface between the architectural task and the imagination of the architect; it is the hand – *the thinking hand*, as Pallasmaa calls it – whose probing action drives the imagination of the architect towards the achievement of the task, the task at hand; it is the hand, in the end, that is the locus of the architect's potency, and for all of us the locus of our human potency.

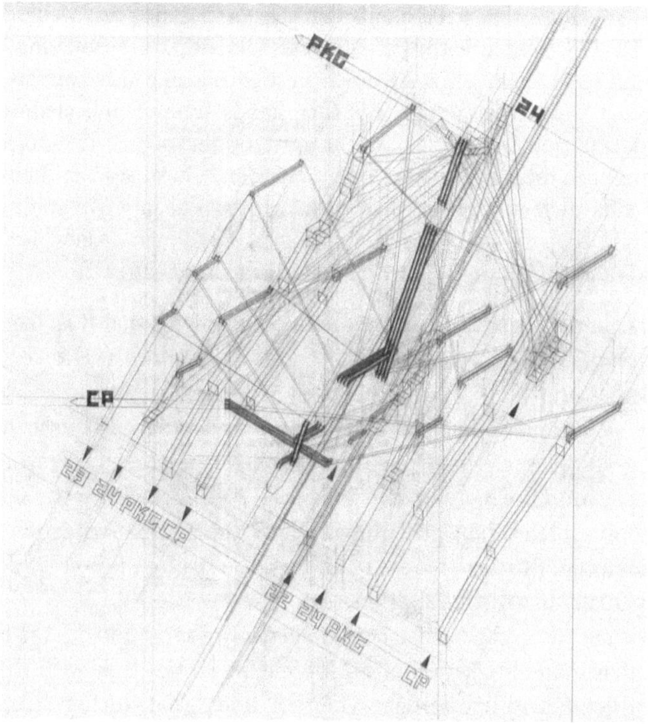

FIGURE 3.3 Composite of circulation paths for a theoretical subway station. P. Delahoy.

The architect's task, Pallasmaa tells us, is "something to be lived, rather than understood"; and it is achieved not rationally-conceptually, but *bodily*, from "unconceptualised and lived existential knowledge" (p. 15). Architectural ideas are thus, for Pallasmaa, biological ideas, and architectural pedagogy is a transfer of a kind of skilled sensation "from the muscles of the teacher to the muscles of the student" through a kind of manual mimesis.

The hand is thus, for Pallasmaa, the periphery of the self; the place where subjectivity pours outwards into the world. But it is also the place where the world forces itself inward, forces itself into subjectivity through the hand, inflecting and modulating that subjectivity through manual action. Pallasmaa begins his book with this passage from Rilke, and the metaphor of the *delta*:

> There are hands that walk, hands that sleep and hands that wake; criminal hands weighted with the past, and hands that are tired and want nothing more, hands that lie down in a corner like sick animals who know no one can help them. But then hands are a complicated organism, a delta from which life from the most distant sources flows together, surging into the great current of action. Hands have stories; they even have their own culture and their own particular beauty. We grant them the right to have their own development, their own wishes, feelings, moods and occupations. (cited in Pallasmaa, pp. 28–29)

For Pallasmaa, then, the essence of manual agency is that it is, like a delta, *bidirectional.* Manual acts exercise so-called *illocutionary force:*[19] so the hand blesses, the hand refuses, the hand expresses friendship and love, the hand demeans and abases – as Pallasmaa tells us, there is a vast theatre of manual gestures, such that when they are performed they find resolution in domains beyond the hand, that is, beyond the gesture. But at the same time the hand modulates the world's inward flow, the world's impact on the self, in a "softening of the existential boundary, the fusion of the world and the self," especially in artistic experience (p. 19).

And what of reason? Pallasmaa makes an observation that is reminiscent of that made by Catton and Montelle in connection with the acquisition of geometric knowledge, when he notes that during creative acts of architecture, a similar suppression takes place, of reason in its analytic mode; the suppression of, in his terminology, "rational-conceptual reason." In the drawings of Alvar Aalto, for example, Pallasmaa notes the "seminal role [in them] of the absent-minded hand and its seemingly unconscious and aimless play" in subconscious sketches of mountain landscapes, and of "many suns" (p. 73).

To the question "what *is* the hand?," Pallasmaa tells us:

> Everyday use of the word as well as classical *surface anatomy* would probably argue that the hand is the organ that extends from the wrist to the fingertips. From the point of view of *biomechanical anatomy*, the human hand would be seen as an integral part of the entire arm. But the arm also functions in a dynamic coordination with the muscles of the neck, back,

[19] Austin's theory of speech acts develops the idea that an utterance can have so-called illocutionary force, as in the utterance, "I declare thee man and wife," which, when it is said, by an authorised person effectuates a legal marriage. Put simply, an illocutionary act is the act performed in making the utterance, as opposed to the act of making the utterance. See Austin (1955/1962), *How to Do Things with Words.*

> and even the legs, and in fact with the rest of the body ... When I raise my hand for an oath or greeting, or give my fingerprints as evidence of my identity, the hand stands for my entire persona. ... Altogether, we are bound to admit that the hand is everywhere in our body, as well as in all our actions and thoughts, and thus the hand is fundamentally beyond definability. (pp. 30–31)

Thus, the mind-hand barrier, following Pallasmaa's thought here, is porous, or at the end of the day, no barrier at all – the hand is everywhere in our body and in our, thoughts.

Note the admonition of Catton and Montelle, that if one is to "do geometry with the Greeks", one must "move about" the Euclidean diagram as one constructs it. Pallasmaa takes us from "moving about" to full habitation:

> While drawing, a mature designer and architect is not focused on the lines of the drawing, as he is envisioning the object itself, and in his mind holding the object in his hand or occupying the space being designed. During the design process, the architect *occupies the very structure that the lines of the drawing represent.* As a consequence of the mental transfer from the actuality of the drawing or the model to the material reality of the project, the images with which the designer advances are not mere visual renderings; they constitute a fully haptic and multi-sensory reality of imagination. The architect moves about freely in the imagined structure, however large and complex it may be, as if walking in a building and touching all its surfaces and sensing their materiality and texture. This is an intimacy that is surely difficult, if not impossible, to simulate through computer-aided means of modelling and simulation. (p. 59; emphasis added)

Pallasmaa's architect is equipped with a fully embodied, *complete* capability, one that fuses into a *single* entity hand, imagination, diagram and task. He lays out a potent confrontation between the architect and the architectural task, in which the architect transports himself into the drawing, and beyond that into the structure the drawing represents, as if the architectural task were like a dream in the mind of the architect, but one that is fully veridical.

Architectural practice has not otherized the body in its critical discourse, as the epistemology of mathematics has done. Pallasmaa's architect is not a metaphysician but *homo faber*: the creator of a public world that shapes and structures every aspect of human existence. As such, Pallasmaa's architect is fully grounded, a maker of his own fortune: *homo faber suae quisque fortunae* – an inhabitant, in other words, of drawings.

THIRD MOMENT: THE INCIDENTAL DRAWING

The genre of the mathematical drawing admits, as we noted, a second form, what I have called the *incidental* drawing (e.g. Figure 3.4). These drawings are not proofs, though that claim has been made for them on occasion; and they are almost always discarded – in fact, it seems wrong to preserve them, as if to circumscribe an epiphany, an intimate, rational moment, with a souvenir. However, we do show them in this volume, "pivotal sources of mathematical meaning" (de Freitas & Sinclair, 2012, p. 137) as they are. Treated as artworks – and in the present environment thinking of these drawings as artworks is completely unproblematic – there is a hiddenness about them, a sense of inwardness, and their aesthetic power, if one wants to think of it that way, seems to come from their hiddenness. As works of art, they sometimes give us *the beauty of the awkward line.* They are meant to depict mathematical gestalts in as simple and unaffected a way as possible, but, aesthetically speaking at least, their exact properties, as Manders calls them, also seem to matter.

FIGURE 3.4 Boris Zilber at his blackboard.

Perhaps one can think of such drawings as a means of controlling the territory, a border-crossing device that carries lived sensibility – the arc of a projectile, the flow of a river, the bodily experience of space and time – into the mathematician's exact discourse. In that sense what the drawings help us to see is that mathematical practice is essentially *bidirectional*; that the mathematician's condition, as it were, is one of ambivalence: the mathematician looks forward, towards exactness, rigor, algebraization, and formalisation in an exact, formal language; but the mathematician also looks backward, towards raw meaning and lived sensibility.

Here too the incidental mathematical drawing helps us to devise better ways of considering the traditional categories of truth and proof, syntax and semantics. I referred earlier to the distinction between meaning and logical form, and the idea that the syntax/semantics distinction emerged in the late nineteenth century on the heels of a long discourse in mathematics centred on diagrammatic inference, a discourse that culminated with Pasch's suppression of the Euclidean diagram in 1882.[20]

Specialized to inference the syntax/semantics distinction comes to us through Bolzano, among others, who distinguished contentual, world-involving inference from mere logical or tautological inference, as we saw,[21] though the notion would be fully articulated only later, by Tarski (1983) in his 1933 paper, and completely articulated by Tarski and Vaught (1958). Specialised to geometry, the distinction comes down to us through Pasch, "the father of rigour in geometry," (Freudenthal, 1962, p. 619) who demanded of a rigorous geometrical discourse, that it does away with meanings: "During the deduction it is useful and legitimate, but in *no* way necessary, to think of the meanings of the terms; in fact, if it is necessary to do so, the inadequacy of the proof is made manifest" (Pasch, 1882; cited in Nagel, 1939, p. 197; emphasis in the original).

The model-theoretic conception of language, as Danielle Macbeth calls it, the fruit of the tradition that begins with the suppression of meaning in the nineteenth century, partitions mathematical discourse, or as Macbeth would have it, burdens mathematical discourse, with a dual structure: there is deductive inference, conceived of as a formal, uninterpreted calculus, a

20 Pasch's move took place not in isolation but in the context of the development of rigor in science altogether. Daston and Galison (1992) trace objectivist imperatives in science overall to the 1830s, when, compelled by what they claim was a growing fear of subjectivity in science, "noninterventionist" or "mechanical" objectivity (p. 82) began to suppress contentual discourse, and by extension, theories that would involve, through an injudicious and uncareful use of words, the *self.*

21 See Bolzano (1992).

mechanical manipulation of empty signs according to syntactic rules; and then there are domains of interpretation, set-theoretical objects that interpret the deductive formalism, and more than that, licence talk of truth.

As a self-account for the mathematical logician this dual structure has turned out to be very useful. But a century of foundations has taught us that splitting meaning off from language at the same time fails to capture the sense of the practice itself, which is contentual and meaningful. Perhaps the problem of developing contentual inference – putting the fly back into the fly-bottle, as it were[22] – is part of the reason why contemporary philosophy of mathematics has taken a turn towards the practice.

The message of the incidental drawing underscores this move towards the practice. The careless use of words and, for that matter, diagrams; the lack of attention to their unintended meanings, can – and has (for the concept of continuity, famously) – introduced instability into mathematics. This made urgent the formulation of a dual structure, one consisting of a precise inference expressed in a formal logical signature, in which mathematical reasoning is identified with an uninterpreted logical calculus; and secondly the concept of a domain of interpretation, in which meaning is handled by severing it from inference, relegating it to the idea of what holds of a concept in such a domain.

The mathematician (who is not a logician) has never internalised this dual structure, despite a century of progress in the foundations of mathematics. Unlike the logician, the mathematician grounds herself in natural language – *and in the production of images.*

So in thinking about the question of the incidental drawing or, beyond that, the question whether a mathematical idea – or say, a proof – can be treated as an artwork – and the kinship that really matters here is of the incidental drawing with the artwork – one has to start by thinking about the strange morphologies of natural language in mathematics; how these morphologies multiply themselves across the mathematical field, while at the same time they can also become submerged and hidden, even for logicians, whose mathematical lives begin and end with all of the agitations and confusions of mathematical languages.

CONCLUSION

I began this essay with Euclidean diagrammatic practice, and the epistemological problems – the philosophical opportunities – raised by it. My

[22] paraphrasing Wittgenstein in Philosophical Investigations: "a philosophical problem has the form: 'I don't know my way about.'" (PI 123), and hence the aim of philosophy is "to show the fly the way out of the fly-bottle" (PI 309)

suggestion was to resolve the problem through the articulation of the notion of the imagination, seen as a manifold enabling the operation of synthetic reasoning. The suggestion did not gain traction, in the light of the rigorization of mathematics in the nineteenth century, which would lead mathematicians away from the diagram, and the idea of sense, or as we would say nowadays, from mathematics conceived purely semantically – not from the point of view of the practice, but from the point of view of its foundations.

What remained unsolved was the problem of manual-practical agency, the problem of articulating a notion of inference harmonizing with it. I looked for a resolution in the single text point of view, based on the immediate assimilation of a visual mathematical. I then looked to architectural practice – the architect's power, the capability he has, to occupy the very structure that the drawing represents.

For the mathematician too, this is the essence of her capability: the power to occupy a structure, to inhabit the very structure that the drawing represents. As a power of thinking – the power to project oneself into a drawing or, beyond that, into a structure – it is fallible. Mistakes are made; special cases, instances of assumptions that were not made evident by the diagram, turn out to be counterexamples. Like Euclidean drawings, whose wild progeny they are, the incidental drawing reminds us that a stream of lived sensibility underlies mathematical practice – a substrate of meaning that pulls the mathematician towards the direct visual assimilation of content, explaining, perhaps, why fallibility occurs so rarely in mathematics.

I leave this as a suggestion, as I leave to the embodiment theorists, the question of the *embodied mathematical.* The case for embodied cognition has been made amply in the general case, and is beginning to specialise to the epistemology of mathematics. What a welcome development! that the discourse of the Timaeus, the Platonic conception of mathematics as a field of pure concepts that leaves the "meat" of the body behind, is beginning to loosen its grip on us (see Jones & Galison, 1988/2013, p. 12).

ACKNOWLEDGEMENTS

I would like to express my deepest thanks to Nathalie Sinclair for inviting me to contribute to this volume. For stimulating discussions and correspondence on the mathematical diagram I am very grateful to Nathalie, to Andrew Arana, who took the time to share the thinking behind his important paper on the mathematical imagination, to Arno Schubbach and to Philip Welch. I would like to thank the Isaac Newton Institute for Mathematical Sciences, Cambridge, for support and hospitality during the programme 'Mathematical, Foundational and Computational Aspects of the Higher Infinite', where work on this paper was undertaken.

REFERENCES

Arana, A. (2016). Imagination in mathematics. In A. Kind (Ed.), *The Routledge handbook of philosophy of imagination* (pp. 463–477). Abingdon, UK: Routledge.

Austin, J. (1955/1962). *How to do things with words*. Oxford: Clarendon Press.

Avigad, J., Dean, E. & Mumma, J. (2009). A formal system for Euclid's Elements. *The Review of Symbolic Logic*, 2(4), 700–768.

Bolzano, B. (1992) *Gesamtausgabe* (Reihe I: Schriften, Band 13,3: Wissenschaftslehre §349–391) (Edited by J. Berg et al.). Stuttgart, Germany: Frommann-Holzboog.

Catton, P. & Montelle, C. (2012). To diagram, to demonstrate: to do, to see, and to judge in Greek geometry. *Philosophia Mathematica*, *20*(1), 25–57.

Coffa, A. (1991). *The semantic tradition from Kant to Carnap: To the Vienna station*. Cambridge: Cambridge University Press.

Courant, R. & Robbins, H. (1941). *What is mathematics? An elementary approach to ideas and methods*. Oxford: Oxford University Press.

Daston, L. & Galison, P. (1992). The image of objectivity. *Representations*, 40, 81–128.

de Freitas, E. & Sinclair, N. (2012). Diagram, gesture, agency: Theorizing embodiment in the mathematics classroom. *Educational Studies in Mathematics*, *80*(1/2), 133–152.

Foucault, M. (1966/2009). *The order of things: An archaeology of the human sciences*. Abingdon, UK: Routledge.

Franks, C. (2014). Logical completeness, form, and content: An archaeology. In J. Kennedy (Ed.), *Interpreting Gödel: Critical essays* (pp. 78–106). Cambridge: Cambridge University Press.

Freudenthal, H. (1962). The main trends of the foundations of geometry in the 19th century. In E. Nagel, P. Suppes & A. Tarski (Eds), *Logic, methodology and philosophy of science* (pp. 613–621). Palo Alto, CA: Stanford University Press.

Giaquinto, M. (2015). The epistemology of visual thinking in mathematics. In E. Zalta (Ed.), *The Stanford Encyclopedia of Philosophy*. (https://plato.stanford.edu/archives/win2015/entries/epistemology-visual-thinking/)

Hilbert, D. (1899/2004). Die Grundlagen der Geometrie. In M. Hallett & U. Majer (Eds), *David Hilbert's lectures on the foundations of geometry 1891–1902* (pp. 436–525). New York: Springer.

Hilbert, D. & Cohn-Vossen, S. (1932/1952). *Geometry and the imagination* (trans. P. Neméyni). New York: Chelsea Publishing Co.

James, W. (1978). *Pragmatism: A new name for some old ways of thinking*. Cambridge, MA and London: Harvard University Press.

Jones, C. & Galison, P. (1998/2013). Introduction: Picturing science, producing art. In C. Jones & P. Galison (Eds), *Picturing science, producing art* (pp. 1–23). Abingdon, UK: Routledge.

Macbeth, D. (2014). *Realizing reason: A narrative of truth and knowing*. Oxford: Oxford University Press.
Manders, K. (1995/2008). The Euclidean diagram. In P. Mancosu (Ed.), *The philosophy of mathematical practice* (pp. 80–133). Oxford: Oxford University Press.
Nagel, E. (1939). *The formation of modern conceptions of formal logic in the development of geometry*. Osiris, 7, 142–223.
Nemirovsky, R. & Ferrara, F. (2009). Mathematical imagination and embodied cognition. *Educational Studies in Mathematics 70*(2), 159–174.
Netz, R. (1998). Greek mathematical diagrams: Their use and their meaning. *For the Learning of Mathematics, 18*(3), 33–40.
Netz, R. (1999). *The shaping of deduction in Greek mathematics: A study of cognitive history*. Cambridge: Cambridge University Press.
Pallasmaa, J. (2009). *The thinking hand: Existential and embodied wisdom in architecture*. Chichester, UK: John Wiley & Sons.
Pasch, M. (1882). *Vorlesungen über neuere Geometrie*. Leipzig, Germany: Teubner.
Quine, W. (1951/1996). Two dogmas of empiricism. In A. Martinich (Eds), *The philosophy of language* (3rd edn, pp. 39–52). Oxford: Oxford University Press.
Roth, W.-M. (2010). Incarnation: Radicalizing the embodiment of mathematics. *For the Learning of Mathematics, 30*(2), 8–17.
Russ, S. (1980). A translation of Bolzano's paper on the intermediate value theorem. *Historia Mathematica, 7*(2), 156–185.
Schlimm, D. (2010). Pasch's philosophy of mathematics. *The Review of Symbolic Logic, 3*(1), 93–118.
Seidenberg, A. (1975). Did Euclid's Elements, Book I, develop geometry axiomatically? *Archive for History of Exact Sciences, 14*(4), 263–295.
Shin, S. (1994). *The logical status of diagrams*. Cambridge: Cambridge University Press.
Shin, S., Lemon, O. & Mumma, J. (2014). Diagrams. In E. Zalta (Ed.), *The Stanford Encyclopedia of Philosophy*. (https://plato.stanford.edu/archives/win2014/entries/diagrams/)
Tarski, A. (1983, 2nd edn). *Logic, semantics, metamathematics: Papers from 1923 to 1938*. Indianapolis, IN: Hackett Publishing Co.
Tarski, A. & Vaught, R. (1958). Arithmetical extensions of relational systems. *Compositio Mathematica*, 13, 81–102.

4

Concepts as Generative Devices

ELIZABETH DE FREITAS AND NATHALIE SINCLAIR

In this chapter, we make six speculative claims about mathematical concepts. Inspired by the work of the historian and philosopher of mathematics Gilles Châtelet, we emphasize the flexibility and generative potential of mathematical concepts, and we focus on the tension between the logical and ontological dimensions of mathematical activity. In the spirit of playfulness, we structure our chapter as a series of propositions.

PROPOSITION 1: CONCEPTS ARE NOT MERELY METAPHORS OR REPRESENTATIONS

Theories of embodied cognition have argued that the sensori-motor experiences of humans – which are said to occur within the physical world – enable humans to create, understand and learn concepts like number. But such theories say next to nothing about what numbers are, let alone how mathematical concepts such as number can inhabit the physical world rather than some detached ideal realm. Trying to understand how seemingly abstract concepts become parts of body-assemblages does not simply involve locating the concrete sensori-motor activities that supposedly give rise to mathematical concepts *as metaphors*. When Lakoff and Núñez (2000) describe the container metaphor from which the mathematical idea of 'set' emerges, they treat concepts as metaphorical *representations* of the real world. If concepts are only metaphorical in relation to the 'real', then we are forced to wonder how and why that metaphorical relationship holds. Metaphoric relationships operate according to a double-standard ontology, where the mathematically abstract and the physically concrete are mutually detached and then reconnected through analogy or resemblance. Such an approach tends to treat concepts as *representations* rather than material arrangements, and it is this material arrangement in which we are most interested.

PROPOSITION 2: CONCEPTS ARE NOT MENTAL CONSTRUCTS ABSTRACTED FROM THE MATERIAL WORLD

Within many contemporary learning theories, concepts are assumed to be abstracted (by a human subject) from a series of encounters with material objects. This approach aligns with the empiricist tradition, and finds expression in various pedagogies that advocate for building student capacity to generalise inductively from such encounters. For example, the concept of square will be abstracted from a series of encounters with material 'squares'; these encounters will be primarily visual and haptic where the child plays with the objects or signs, each an instantiation of the concept, until the child is adequately familiar and able to identify what all of these instances have in common, that being the concept of squareness. Similarly, the concept of '5' will be abstracted from a series of encounters with material sets of five counters or cups or blocks, which may be touched, pointed at and counted or even subitised (visually apprehended). This theory of abstraction informs much of our work in education, and yet there are tacit assumptions in this approach that need to be examined for how they fail to address important aspects of learning. The limitations of this approach, which critics have voiced for millennia, pertain to various issues. For example, this approach suggests that concepts are abstracted away from the physical situation, and that the learner is granted only enough embodiment to perform that abstraction. Miraculously, the material sensation becomes immaterial (a dualism of mind and body). This approach remains committed to an abstract/concrete binary that ultimately abandons mathematics to a realm of the inert and disembodied.

Instead, we want to pursue the proposal that mathematical concepts are material in much the same way that Karen Barad (2007) invites us to think of atoms, photons, stingrays and lightning – as unfinished relational assemblages or arrangements. We claim that there is no *definitive detachment* from the encounter with the square-shaped tile or the set of five counters. In order to explore the material and embodied aspects of learning, we need to rethink abstraction as a process of creative speculation. Our concepts – of squareness, fiveness, etc. – thus retain the trace of the movement of the eye, hand and brain. In retaining this mobility, as

Châtelet (1993/2000) points out, our proposal differs significantly from other explanations of mathematical concepts that appeal to conventional 'abstraction', which is often contemptuous of the hand and the movement by which the possible was determined or actualized. Abstraction strips mathematical entities of their physical emergence and pretends that they have no experiential history. Although many scholars assume that it is this act of stripping away the contingencies of emergence that makes mathematics so effective in science, Châtelet's entire historical project is to contest this assumption and show how inventions in mathematics proved effective *precisely* because they retained a trace of the gestures (and the mobility of the material) that brought them forth. Thus, concepts in sensible matter become intelligible, not by a *reductionist abstraction* or by a "subtraction of determinations" (Aristotle's approach to abstraction), but by the actions (moving, excising, cutting through diagrams and gestures) that awaken potential multiplicities that are always implicit in any material.

PROPOSITION 3: CONCEPTS ARE VIBRANT AND INDETERMINATE, HAVING ONE FOOT IN THE VIRTUAL AND ONE IN THE ACTUAL

The *virtual* has a long philosophical lineage. Châtelet uses it as an ontological category that enables him to describe multiplicities that are real but that are never fully formed, determined or exhausted, and thus never actualised. A *multiplicity* contrasts with an *essence* in that the latter has a uniform and timeless identity, while the former is divergent and alters historically. In conventional usage, if we take an object to have an essence, then we assume that it is indifferent to its external forms. On the other hand, to say that an object expresses a multiplicity is to say that matter is spontaneously capable of generating pattern, and that such pattern does *not resemble* some pre-given essence. Thus, the virtual is that which is latent in matter – but what is latent is not essence, ideal or form; rather, it is mobility, vibration, potentiality and indeterminacy.

We draw on Châtelet's notion of the virtual, which he applies to mathematics, and which he uses to resolve the 'paradox' that arises (from Plato) when the mathematical and the physical are taken to be ontologically distinct. The paradox concerns the impossibility of a *causal* relationship between the mathematical (as that which pertains to eternal truths) and the physical (as that which pertains to the contingent and the real). Plato's solution was to posit a timeless being that 'recollects' the ideal forms of mathematics. Aristotle's solution was to infer the existence of some 'final cause' that can ensure cohesion between the two natures.[1] Galileo courageously proposed an alternative to Aristotle's teleological explanation, instead suggesting a link between the *genesis* of mathematical and physical concepts. What could this link be? Châtelet attends to this same paradox, through a *monist ontology*, claiming that the *virtual* provides the means to overcome the Aristotelian embargo against the mobility of mathematics.

Thus, for Châtelet, mathematical concepts partake of the material, insofar as they are both actual and virtual, the two dimensions of the real. This might be difficult to imagine with concepts such as fiveness and squareness, which seem fixed, universal and eternal. And indeed, Châtelet's analyses involve historically recent mathematical concepts such as the complex number and the point at infinity, which have yet to be tamed. But there are ways in which 'simple' concepts like number are embedded in or animated by these more complex concepts. The mathematical continuum, for instance, is a concept that undergirds concepts of number. The mathematical continuum refers to both the geometric number line and the real number system that occupies it. Concerns that Euclid's axioms could not, in principle, construct the continuity of the number line led to various attempts to do so in the nineteenth century. Dedekind (1831–1916), intent on banishing all geometric 'intuition' from mathematics, used sets and 'cuts' to compose the infinite granularity needed for the continuum. Cantor (1845–1918) would offer a similar approach, proposing necessary and sufficient conditions for continuity that relied on set-theoretic constraints. These attempts to erase the materiality of the number line reveal an awkward haunting. How can a line be composed of points? How can number – no matter how numerous – form a continuum? The mathematical continuum seems to vibrate with the virtual, with the potentiality of the discrete and continuous, the counted and the uncountable, the separate and the connected.

The number line in school mathematics is often seen as a fixed extension on which position (the number 5) can be located. Drawing on the traumatic history of the continuum, we can disrupt this image of the number line (and the numbers that are located on it) by rethinking it as somehow made muscular through the mobility of matter. If motion is the ontogenetic force by which position comes into being, then 5 could be the starting point for a new vertical extension that cuts the number

[1] As Châtelet describes, Aristotle later sketched out another solution to the problem, which Leibniz studied and eventually developed along the lines described in this chapter.

line, or it can be the genesis of a fold. Indeed, Châtelet uses the 'virtual' to describe the 'indeterminate dimension' in matter that literally destabilizes the rigidity of extension (p. 20). He describes these elastic folds in terms of transverse and vertical impulses that push through the apparent rigidity of extension.

In rethinking the number line in schools, we might want to recast it as elastic, so that students can carve out the virtual real numbers embedded between whole numbers by grabbing and stretching the number line in order to bring forth an infinitude of numbers that were imperceptible a moment earlier. Students might then explore the density of the number line through this engagement – an engagement made all the more embodied when we see how an elastic line, more so than a rigid one, seems to sustain the potential of our flowing gestures. The number line becomes a quivering space of potential rather than an interval contained between two fixed points. The number line becomes a highly animate concept made vibrant and creative through the indeterminacy (virtuality) buried in it.

Thus, we believe that mathematics cannot be divorced from matter, and it is the virtual dimension of this matter that animates the mathematical concept. Mathematicians engage with this dimension in thought experiments, but these are not the disembodied mental ruminations with which we typically associate mathematical thinking; rather, they are physical experiments in which objects are set in motion, cut out from the plane, plunged into singularities, stretched around points of inflection, packed into Abelian groups and so on. When geometric points are seen as mobile material bodies, the mathematician can direct them to new places and even lift them off the Cartesian plane so that they engender new kinds of entities in the complex field. In other words, mathematical concepts engage in a process of *becoming*, a process that involves human and other kinds of material activity.

Because of this process of becoming, a concept is never identical to itself, and thus it is not a *form* that is somehow buried *fait accompli* in matter and waiting to be conjured or evoked. The virtual is not an *a priori* invisible form, and it is not the case that ideal mathematical entities find their inaccurate, fuzzy realizations in the real world. The virtual is not a realm of forms to be mirrored in the physical world. *The virtual does not resemble the actual.* This lack of resemblance is crucial in appreciating the power of the virtual to disrupt the common Platonist epistemology about mathematics.

PROPOSITION 4: CONCEPTS OPERATE AS BOTH LOGICAL AND ONTOLOGICAL DEVICES

Châtelet describes the activity of mathematics as entailing two movements, one between the virtual and the actual, and the other between the possible and the real. Mathematical activity thus involves both *actualizing the virtual* and *realizing the possible.* The possible/impossible distinction structures and limits the appearance of the entity according to current rules of inference and perceptual habits (for example, a number can be either even or odd, but not both). The possible pertains to the compliance of our actions with logical and related perceptual constraints. The virtual pertains to the indeterminacy at the source of all actions and all matter. Alternative actualizations might involve carving out a new mathematics in which certain numbers such as 6 and 10 are both even (because they are divisible by 2) *and* odd (because they form an odd number of even groups).

Together, actualizing and realizing bring forth something new into any given situation; but *realizing* plays by the rules of logic (inductive or deductive inferences), while *actualizing* involves a different kind of determination, one that generates something ontologically new, like a complex number or a point at infinity of projective geometry – neither of which emerges from the rules of logic alone.

It would seem that many mathematical objects can be thought of as functioning both logically and ontologically. The circle as the set of points equidistant from a given centre realizes the possibility of the circle, namely the set of points *determined* by the constraint of the definition. In instantiating or realizing the definition in the circle, we abide by the logical necessity of the meaning of the words in the definition. But there is also another kind of determination, material rather than logical, and this kind of determination involves an actualizing of the virtual dimensions of matter. The circle is *also* materially produced by the trace of a point constantly being pulled to a centre by an equal reciprocal force as it moves forward. In other words, there is a mobility inherent within the circle, and thus a way of conceiving of the very same circle through this mobile generative activity. The definition of a locus of points performs the logical necessity of the circle, the seeming acontextual universality of the concept, while the activity of generating or making the circle (by whatever means) is just as important in *determining* its existence. In other words, the concept of circle *lives in* material media as much as it *lives by* logical constraints.

Most if not all mathematical concepts are determined both logically and ontologically. Indeed, the historical relationship between logic and mathematics sheds light on these two aspects of mathematical activity. Hacking (2014) argues that the connection between logic and mathematics 'simply did not exist' until the logicist movement of the nineteenth century (advocated by Frege, above all), which aimed to reduce mathematics to logic, and replaced Aristotelian logic with what was termed 'symbolic logic' (p. 137). Today, symbolic logic is taught as part of geometry courses in US high schools, but, according to Hacking, this curriculum is a 'fluke' historical association. This historical claim contrasts with the myth of Greek mathematics as historically significant for its revolutionary use of proof and demonstration. Hacking points to Kant's role in the propagation of that myth: '[A] new light flashed upon the mind of the first man (to be Thales or some other) who demonstrated the properties of the isosceles triangle' (Kant, cited in Hacking, p. 125). Latour (2008), in agreement with Hacking regarding the mythologizing of Greek proof, emphasizes Plato's role in forging an image of mathematics as the ultimate face of logical deduction, in which argument was seen to be *apodictic* rather than rhetorical and sophistic. Both Hacking and Latour draw on the work of Netz (1999), who shows how the ancient revolution (or 'catastrophe', as he calls it) was a far more applied or practical affair, grounded in material conditions and experiments, rather than appeals to logic.

There were two kinds of mathematics in ancient Greece, the mathematics of elite membership in Plato's academy, where apodictic proof was prized, and the far more populous and influential mathematics of the trade schools (Asper, 2009). Thus, the historical contingency of the later association of mathematics with logic, and the need to reconsider, as Latour and Hacking do, the political reasons for such an association, invites us to consider alternatives. With reference to the multiplicity of styles of reasoning that function effectively in the sciences, Hacking (2014) says of mathematics: 'I have argued above that we could have got on without proof as the gold standard, indeed without proof at all, but then mathematics would have been a very different kind of activity, far more akin to the empirical sciences than as it is practiced today' (p. 142). The very idea that we could have a different mathematics from the one we have today is exciting and, as Hacking suggests, largely depends on alternative ways of conceiving the relationship between logic and mathematics.

Insofar as mathematics aims to be a self-authenticating method of inquiry, logic will always play a pivotal role, and yet the empirical dimension of mathematics always introduces the semantic or ontological factor.

If there are two compelling reasons (or feelings) for why mathematics inspires philosophical reflection – the feeling of necessity achieved through apodictic proof and the feeling of surprise at the 'unreasonable' effectiveness of mathematical applications – it is because there is always this dual investment in the logical and the ontological. Hacking traces two historical trajectories for mathematical proof, one related to Descartes and the other to Leibniz. For Descartes, proof involves deduction from *certain knowledge*, and thus rests on semantics of reference (reference to truth in the world), whereas for Leibniz, proof is a more syntactic affair, a deduction from proposition to proposition without reference to certain knowledge. According to Hacking, we currently suscribe to a Leibnizian image of proof, especially as we turn to computer generated proofs that are, in the long run, indifferent to human knowledge. This, again, is a historical contingency, and reflects a particular interest in computational proof.

Given this contingency, one can examine particular proofs for how concepts are deployed as both logical and ontological devices. For instance, the ancient Greek proof of the irrationality of $\sqrt{2}$ relies extensively on even/odd parity. The concept of parity is leveraged in this famous proof to unleash and make actual the virtual (the irrational number) while operating according to the laws of legitimate inference. Even/odd parity simultaneously operates as a logical instrument, determining the boundary between sense and nonsense by which the possible is realized, and ontologically. The proof both brings something new into existence and attests to the truth of this affirmation. In other words, a concept of this kind is a multi-purpose apparatus[2] put to work in a proof in order to carve out new mathematical entities. The proof *shows how* the concept of even/odd parity engenders new kinds of mathematical objects. The concept marshals both logical constraints and ontological force; new conceptual assemblages emerge – in this case, the concept of irrational number.

This famous proof, and our understanding of it, evokes a long-standing issue in the philosophy of mathematics regarding the status of non-constructive proofs, that is, proofs that assert the existence of something without furnishing that thing. Brouwer and the intuitionists were concerned that such proofs were dangerous and were likely to lead to paradoxes. Our concern, however, is not to banish non-constructivist proofs,

[2] Deleuze uses the word *dispositif*, which is often translated into 'apparatus' or 'device'. Unlike a device, which conjures the idea of a tool, the notion of *dispositif* is described as being a 'tangible, multilinear ensemble' (Deleuze, 1992, p. 159), or a multiplicity.

but instead to focus on the generative aspect of proofs, and the way that mathematical concepts can and do function as generative devices.

We imagine concepts as apparatus that actualize the virtual. Our concern, in relation to school mathematics, will thus be to resist the downgrading of any list of concepts to a set of curricular topics. This will require *sustaining the concept as a creative force* by ensuring that the concept is put to work in engendering new mathematical entities. Perhaps the way to ensure that the concept remains operative, mobile and creative is to focus on whether the work actually brings the new into existence. Such work would then tap into or plug into the virtual dimension of matter. When the concept is used only as a logical tool, or as an object or relation with a finite set of properties, the ontological aspect (actualizing the virtual) is abandoned, and activity is reduced to adhering either to naming exercises (identifying instances of the concept, as discussed in Proposition 3 earlier in the chapter) or to applications of the concept as a rule or logical constraint. In such cases, the ossified concept fails to sustain the mobility and potentiality which is its nature.

PROPOSITION 5: THERE IS NO *A PRIORI* LOGICAL ORDERING BETWEEN MATHEMATICAL CONCEPTS

This claim relates to how we structure curriculum and learning experiences. Since Piaget, the mathematics curriculum has been heavily influenced by what Lundin (2011) calls a 'developmental ideology', namely a rationale based on theories about the order and age at which children (can/should) develop knowledge of particular mathematical concepts. This ideology assumes that learning is a problem of construction and centrally involves the acquisition of logical reasoning skills. With it comes the ensuing metaphor that suggests starting from the ground up and making sure the foundations are solid. Within the ideology of development, the discipline stays fixed and immobile, while the child moves slowly up, forming 'concepts' along the way.

In mathematics education research, the word 'concept' has been associated with the notion of a schema (Skemp, 1979), which is described as a 'mental entity' (p. 114), as well as a 'network of connected concepts' (Olive & Steffe, 2002, p. 109). Etymologically, concepts are things that can be taken in (the mind). Vygotsky (1934/1986) describes concepts more functionally as 'taking shape in the course of a complex operation aimed at the solution of some problem' and insists that a concept 'is not an isolated, ossified, and changeless formation, but an active part of the intellectual process, constantly engaged in service of communication, understanding and problem solving' (p. 98).

At a basic level, the scope and sequence metaphor for curriculum involves an ordered list of concepts, which is often taken to be the 'natural' order that should be followed when moving on to higher-level concepts without leaving any holes or gaps. For example, in North American curricula, the 'natural' order dictates that addition should come before subtraction, which serves to differentiate these two operations as two distinct (albeit related) concepts. However, in China, addition and subtraction are taught together, as a unity of three numbers *a*, *b* and *c*, in which $a + b = c$ *and* $c - b = a$. This might suggest that addition and subtraction are part of the same concept. This difference in curriculum underscores the historical contingency of concepts, and troubles the developmental approach.

Beyond the basic scope and sequence metaphor, Dietiker (2012) writes that the mathematics education literature offers three additional metaphors of curriculum: *curriculum map*, *learning trajectory* and *story*. The first was used by Dewey (1902) to distinguish between the 'logical' organisation of content and the everyday, 'psychological' experiences of the learner. This metaphor encompasses more than scope and sequence in the sense that it invites reflection on how the parts of the terrain relate to each other, including the particular students involved. Like scope and sequence, it defines concepts through ordering but recognises the way in which concepts can be defined in response to the needs of particular learners (maybe the order of the concepts changes or, more likely, the degree of generality).

A related metaphor of curriculum is that of the *learning trajectory* (Breyfogle, Roth, McDuffie & Wohlhuter, 2010), which focuses on the critical moments along the path that mark progress towards a future goal. A related, student-oriented construct of *hypothetical learning trajectory* (Clements & Sarama, 2004; Confrey, Maloney, Nguyen, Mojica & Myers, 2009) has also emerged, which describes increasing levels of sophistication demonstrated by students as they work with a concept during clinical interviews and teaching experiments. While Confrey et al. (2009) acknowledge that there is not just one trajectory through the 'conceptual corridor', the metaphor of trajectory itself evokes a single, vectored direction towards a fixed node, which is presumably the attainment of a full-blown concept.

A powerful alternative approach to the role of concepts in mathematical activity can be found in the fallibilism articulated by Imre Lakatos in the early 1960s. According to Lakatos, mathematical activity is primarily about concept formation and deformation. Mathematicians are always working with ill-defined or ambiguous concepts, and as they try to prove things about those concepts, they stretch them in new directions. In his famous example, the concept of polyhedron is stretched and moulded, and indeed transformed beyond recognition. However, this perturbation does not occur in a vacuum; it emerges in relation to a particular problem – which, in Lakatos' case, is about the relation between polyhedral vertices, edges and faces. With Lakatos, as with Vygotsky, concepts emerge out of particular problems and proofs, so that one can hardly talk about concepts without also talking about problems. This way of thinking about concept is much closer to that of Châtelet, not only in its connection to problems, but also in its intrinsic focus on what the concept is *doing*, that is, on the mathematical event in which the concept is activated.

PROPOSITION 6: CONCEPTS EMERGE FROM AESTHETICO-POLITICAL ACTS

New concepts arise out of boundary-making practices that often (or always?) begin with some kind of different sharing of the sensible, where sensible refers to what is visible, auditory, intelligible. We say that concepts are aesthetic insofar as they pertain to the senses (what is visible, auditory, intelligible), but also because they necessarily involve questions of values and judgements in terms of what 'makes sense'. In other words, concepts live through particular alignments between sense and sensibility; new concepts disturb those alignments and produce a kind of *dissensus*, which is often temporary, until a new consensus is established. Examples from history abound: zero, imaginary and complex numbers, functions that are everywhere continuous and nowhere differentiable, 17-dimensional spaces, etc. Examples in classrooms also abound, where students, if invited, participate in divergent concept formation. Teachers are often called on to correct these concepts, but it is worth exploring these acts of dissensus as concept-deformation *à la* Lakatos. For instance, in de Freitas and Sinclair (2014), we discuss video data from a grade three classroom taught by Deborah Ball in which the students are trying to decide which numbers are even and which are odd. One student, Sean, proposes that 6 could be *both* even and odd. Several students try to convince him otherwise, each using a different form of 'common sense': these students state that even and odd should alternate, as made visible by a student who taps a wooden pointer on the image of a number line glued to the wall; that even numbers should be shared evenly, as made visible by a student who comes to the blackboard and draws the splitting of six circles into two even parts; that even numbers should have one definition, and odd another, as suggested by the teacher. Sean disrupts these various sense-makings by proposing to partition numbers in a non-binary, non-iterative way – numbers can be both even and odd! His proposal is completely plausible and logically coherent, and he argues convincingly for it. Thus, Sean proposes a new concept of number. We can recognise Sean's 'even and odd' numbers as the even numbers that are *not* powers of two. But the important insight is that Sean's proposal is a deformation of a concept, a highly valued activity in mathematics, which calls into question what will be taken to 'make sense' in this mathematics classroom.

CONCLUDING COMMENTS

Based on these six claims, we reclaim the concept as a creative material force in the experimental tradition, shifting our attention to the onto-generative nature of mathematical activity. We hope such an approach might rescue mathematical concepts from the staid curricular lists which entomb them, and instead reanimate them as ontogenetic devices. We are not advocating for concepts as transcendent universal ideals, but as operative and highly flexible arrangements or apparatus *in this world*. A concept brings with it an entire 'plane of immanence' on which we operate, explore and create. Concepts are entirely immanent within the real, which is what makes them such powerful devices for carving up the real. We suspect that it is the very indeterminacy and mobility of concepts (their virtual dimension) that grants them their creative power. This mobility is precisely what Châtelet reveals in the various historical examples he considers. In a compelling inversion of the usual aim for learning, Cutler and MacKenzie (2011) suggest that *learning is that which sustains the mobility of concepts* – in other words, 'that which resists determination as knowledge' (p. 68). This chapter has pursued this endeavour by emphasizing the *ontological* force of concepts, not as transcendental ideals that shape reality, but as immanent forces at play in particular problems, operating alongside the logical constraints of mathematical proof. Our aim has been to explore the mutual entailment of matter and meaning in mathematical activity, and to promote awareness of the vitality and indeterminism that lie at the source of curricular concepts.

REFERENCES

Asper, M. (2009). The two cultures of mathematics in ancient Greece. In E. Robson and J. Stedall (Eds.), *The Oxford handbook for the history of mathematics* (pp. 107–132). Oxford: Oxford University Press.

Barad, K. (2007). *Meeting the universe halfway: Quantum physics and the entanglement of matter and meaning.* Durham, NC: Duke University Press.

Breyfogle, M. L., Roth, W.-M., McDuffie, A. & Wohlhuter, K. A. (2010). Developing curricular reasoning for grades preK-12 mathematics instruction. In B. Reys, R. E. Reys, & R. Rubenstein (Eds.), Mathematics curriculum: Issues, trends, and future directions. *NCTM 2010 Yearbook* (pp. 307–320). Reston, VA: National Council of Teachers of Mathematics.

Châtelet, G. (1993/2000). *Les enjeux du mobile.* Paris: Seuil. (Engl. transl., by R. Shore & M. Zagha, *Figuring space: Philosophy, mathematics and physics.* Dordrecht: Kluwer Academy Press, 2000.)

Clements, D. H. & Sarama, J. (Eds.) (2004). Hypothetical learning trajectories [Special issue]. *Mathematical Thinking and Learning*, 6(2), 81–260.

Confrey, J., Maloney, A., Nguyen, K., Mojica, G. & Myers, M. (2009). Equipartitioning/splitting as a foundation of rational number reasoning. In M. Tzekaki, M. Kaldrimidou & C. Sakonidis (Eds.), *Proceedings of the 33rd Conference of the International Group for the Psychology of Mathematics Education* (Vol. 1, pp. 345–352). Thessaloniki, Greece: PME.

Cutler, A. & MacKenzie, I. (2011). Bodies of learning. In L. Guillaume & J. Hughes (Eds.), *Deleuze and the body* (pp. 53–72). Edinburgh: Edinburgh University Press.

de Freitas, E. & Sinclair, N. (2014). *Mathematics and the body: Material entanglements in the classroom.* New York: Cambridge University Press.

Deleuze, G. (1992). What is a dispositif? in T. J. Armstrong (Ed.), *Michel Foucault Philosopher* (pp. 159–168). Hemel Hempstead, UK: Harvester Wheatsheaf.

Dewey, J. (1902). The child and the curriculum. In L. A. Hickman & T. M. Alexander (Eds.), *The Essential Dewey* (Vol. 1, pp. 236–245). Bloomington, IN: Indiana University Press.

Dietiker, L. (2012). *The mathematics textbook as story: A literary approach to interrogating mathematics curriculum.* Unpublished PhD dissertation. East Lansing, MI: Michigan State University.

Hacking, I. (2014). *Why is there philosophy of mathematics at all?* New York: Cambridge University Press.

Lakoff, G. & Núñez, R. (2000). *Where mathematics comes from: How the embodied mind brings mathematics into being.* New York: Basic Books.

Latour, B. (2008). The Netz-works of Greek deductions. *Social Studies of Science*, 38(3), 441–459.

Lundin, S. (2011). Hating school, loving mathematics: On the ideological function of critique and reform in mathematics education. *Educational Studies in Mathematics*, 80(1), 73–85.

Netz, R. (1999). *The shaping of deduction in Greek mathematics: A study in cognitive history.* New York: Cambridge University Press.

Olive, J. & Steffe, L. (2002). Schemes, schemas and director systems (an integration of Piagetian scheme theory with Skemp's model of intelligent learning). In D. O. Tall & M. O. J. Thomas (Eds.), *Intelligence, learning and understanding in mathematics: A tribute to Richard Skemp* (pp. 97–130). Flaxton, Australia: Post Pressed.

Skemp, R. (1979). *Intelligence, learning and action – A foundation for theory and practice in education*. Chichester, UK: Wiley.

Vygotsky, L. (1934/1986). *Thought and language*. Cambridge, MA: MIT Press.

PART III

5

Bernhard Riemann's Conceptual Mathematics and the Pedagogy of Mathematical Concepts

ARKADY PLOTNITSKY

INTRODUCTION: FROM CONCEPTUAL PHILOSOPHY TO CONCEPTUAL MATHEMATICS

In *What Is Philosophy?*, Gilles Deleuze and Félix Guattari give a double definition of philosophy. First, they define philosophy as the invention of new concepts, in their special sense of the term. Secondly, they define philosophy as "a *pedagogy* of the concept" (Deleuze & Guattari, 1994, p. 12). This paper aims, partially and against Deleuze and Guattari's own grain, to transfer this joint definition to mathematics, by using the work of the mathematician whose ideas had a major impact on Deleuze and Guattari's thinking – Bernhard Riemann. Riemann's mathematical thinking was, I argue, fundamentally defined by the invention of new concepts in Deleuze and Guattari's sense, now applied within the mathematical context rather than the philosophical. This transfer is partial because mathematics cannot be defined only by the invention of mathematical concepts, given the roles of other aspects of mathematical thinking, such as logic and axiomatic reasoning, which Deleuze and Guattari see as most defining in mathematics. Accordingly, it becomes a matter of relative precedence of these components at different junctures, with the invention of concepts having the main role in *creative* mathematical thought. But then, one could argue that philosophy cannot be limited to the invention of new concepts either, and that there, too, it is a matter of relative precedence of different aspects of philosophical thinking at different junctures, with the invention of concepts having the primary role in creative philosophical thought. This transfer is enacted against Deleuze and Guattari's own grain, because they see their concept of concept as pertaining primarily or even exclusively to philosophy, in juxtaposition to mathematics and science, or art. If, however, one adopts the view of mathematics as, first, fundamentally based on the invention of concepts and, second, as pedagogy of the concept, one might ask what are the implications of this view for teaching mathematics. I suggest that these implications are significant and shall sketch them in closing.

My argument is grounded by Riemann's Habilitation lecture, "On Hypotheses That Lie at the Foundations of Geometry" [*Ueber die Hypothesen, welche der Geometrie zu Grunde liegen*], given in 1854 but first published in 1868, two years after Riemann's death (Riemann, 1854). The lecture offered a radical rethinking, based on Riemann's mathematical concept of manifoldness or manifold [*Mannigfaltigkeit*], of space (seen by Riemann, as by others at the time, as physical space) and geometry, as against the preceding history of both, from and before Euclid to the discovery of non-Euclidean geometry by Nikolai Lobachevsky and János Bolyai (around 1830) and beyond. Establishing the *possibility* of non-Euclidean geometry by Lobachevsky and Bolyai was a major event, which had profound implications for mathematics, physics, philosophy, and culture. However, as Riemann's lecture made clear, the pre-Riemannian thinking concerning non-Euclidean geometry did not sufficiently depart from previous (Euclidean) thinking, with a possible exception of ideas of Karl Friedrich Gauss, Riemann's teacher and precursor. In enacting his more radical departure, Riemann followed Gauss' ideas about the curvature of two-dimensional surfaces. The non-Euclidean geometry of Lobachevsky and Bolyai did not figure in Riemann's lecture and, unlike Gauss' work, was not even mentioned there. In particular, the fifth (parallel) postulate of Euclid, central to the history that led to the discovery of non-Euclidean geometry, played no role in Riemann's argument. Riemann's thinking was conceptual rather than axiomatic, as against that of Lobachevsky and Bolyai, and it led him beyond only a single alternative to Euclidean geometry to an uncontainable multiplicity of geometries and an even greater multiplicity of possible spaces, because some of them would not admit geometry. The latter circumstance became important for the development of topology as a mathematical discipline, which, unlike geometry, deals with the *architecture* of a given space, and not measurement in that space (using the term space, as we do now, in the mathematical sense).

The implications for physics were dramatic as well, and Riemann's contribution to physics in the lecture was just as momentous as his contribution to mathematics. Before the discovery of the non-Euclidean geometry, one and only one geometry would be available for a geometrical description of physical space (which is, again, how the term "space" was understood at the time) – if one assumes that space could be described geometrically, which had been a grounding assumption of modern physics from Galileo and Newton on. While Kant's epistemology of phenomena (appearances to the human mind) vs. noumena (things-in-themselves) qualified this assumption, it did not change its essential import for physics. Kant, too, assumed that physical space is Euclidean and three-dimensional, or at least is unlikely to be anything other. The situation fundamentally changed with the discovery of the non-Euclidean geometry and then, even more radically, with Riemann, whose argument, as I said, implied that an infinite number of possible geometries could, in principle, be associated with a physical space. Any such association is a hypothesis (this is what "hypotheses" of Riemann's title referred to), which is subject to testing, verification, and so forth. Thus, we know well local physical geometries, say, in the vicinity of the solar system, but it is much more difficult to be sure concerning the ultimate geometry of the Universe, although the current data suggests that it is, on average, flat (has the curvature zero) as far as we can observe it. It is also possible that there are other Universes with different geometries and topologies. Riemann did not envision this possibility, which arises from considerations concerning the Big Bang origin of the Universe. On the other hand, he argued that the character of physical space, "the reality underlying space," on very small scales was unknown at the time and could only be determined by physics. This character remains unknown.

PLANES OF IMMANENCE AND THE ARCHITECTURE OF CONCEPTS

In *What Is Philosophy?*, Deleuze and Guattari define thought [*la pensée*] as a confrontation between the brain and chaos. On the surface, this view is hardly surprising: much of our thinking may be understood as this type of confrontation. Deleuze and Guattari, however, give a special meaning to this confrontation by virtue of their conception of "thought" as different from merely thinking (understood, say, as mental states and processes) and manifested especially in philosophy, art, and mathematics and science. While unremittingly at war with chaos, thought also works together with chaos, rather than merely protects us against chaos, which is the case with other forms of thinking, in particular, opinion. Deleuze and Guattari see chaos (which they understand in a particular way, explained later in the chapter) not only as an *enemy* but also as a *friend* of *thought*, its greatest friend and its best ally in a yet greater struggle, that against opinion, an enemy "like a sort of 'umbrella' that protects us from chaos." As they say,

> [The] *struggle against chaos* does not take place without an affinity with the enemy, because another struggle develops and takes on more

> importance – the struggle *against* opinion, which claims to protect us from chaos.... [T]he struggle with chaos is only the instrument in a more profound struggle against opinion, for the misfortune of people comes from opinion.... But art, science, and philosophy require more: they cast planes over chaos. These three disciplines are not like religions that invoke dynasties of gods, or the epiphany of a single god, in order to paint the firmament on the umbrella, like the figures of an *Urdoxa* from which opinion stem. Philosophy, science, and art want us to tear open the firmament and plunge into chaos. (Deleuze & Guattari, 1994, p. 202)

They see chaos itself as "defined not so much by its disorder as by the infinite speed with which every form taking shape in it vanishes. It is a void that is not a nothingness but a *virtual*, containing all possible *particles* and drawing out all possible forms, which spring up only to disappear immediately, without consistency or reference, without consequence. Chaos is an infinite speed of birth and disappearance" (Deleuze & Guattari, 1994, p. 118). Philosophy engages with chaos by creating planes of immanence and concepts; art by creating planes of composition, and compositions and affects; and mathematics and science by creating planes of reference, and frames of reference and functions (or other mathematical objects) and (logical) propositions.

This specificity makes these fields different from each other. However, the affinities and relationships among art, science, and philosophy may be just as significant, and reflect a more complex landscape of thought in which these fields are positioned. Here, I address conceptual thought in philosophy and mathematics and science. I argue that creative thought in mathematics and science is defined as much by the invention of new mathematical and scientific concepts as is creative thought in philosophy by the invention of new philosophical concepts.[1]

I shall now explain Deleuze and Guattari's (1994) *concept* of a philosophical concept. A concept is not an entity established by a generalisation from particulars or "any general or abstract idea" (pp. 11–12, 24). A concept is a multicomponent entity, defined by the specific *organization* of its components,

[1] In addition, although the subject can only be mentioned here, mathematics, science, and philosophy may involve planes of composition and the creation of compositions, found in artistic *thought*, while the latter may, conversely, involve planes of immanence and the creation of concepts, or planes of reference. A creation of (new) objects of the mathematical type is unusual in art, which, however, modernist painting, for example, borrows from and sometimes enriches mathematical concepts, as well as serves as an inspiration for the creation of mathematical concepts. Literature may create new logical structures that are in effect those of mathematical logic. Lewis Carroll's work is an obvious example, although it is a special case because he was a mathematician.

which may be general or particular, and some of these components are concepts in turn: "There are no simple concepts. Every concept has components and is defined by them.... It is a multiplicity.... There is no concept with only one component" (p. 16). It is the relational organization of components that is most crucial in defining a concept. A single-component concept is only a product of a provisional cut-off of its multi-component organization, as would, for example, be the concept of "bird," discussed later here, if one disregards its components. In practice, there are always cut-offs in delineating a concept, which results from assuming some of the components of this concept to be primitive entities whose structure is not specified. These primitive concepts could, however, be specified by an alternative delineation, which would lead to a new concept, containing a new set of primitive (unspecified) components. The history of a concept – and every concept has a history – is a history of such successive specifications and changes in previous specifications.

Consider the concept of "bird," even as it is used in our daily life. On the one hand, it is a single generalisation. On the other hand, what makes this concept that of "bird" is the implied presence of components or sub-concepts, such as "wings," "feathers," and "beak," and the relationships among them. The concept of bird acquires further features and components and thus becomes a different concept, in zoology, and still a different concept in biology. A philosophical concept of a tree is yet something else. According to Deleuze and Guattari, "a [philosophical] concept of a bird [one such concept!] is found not in its genus or species but in the composition of its postures, colors, and songs, something indiscernible that is not so much synesthetic as syneidetic" (p. 20) (a product of the synthesis of the eidos, form, of each concept it emerges from or absorbs). Each concept is also defined as a problem, which is as multifaceted as the concept itself is, a definition that has a mathematical genealogy. While it must be solved, a problem does not disappear in its solutions: it is "determined at the same time as it is solved" and is "at once both transcendent and immanent in relation to its solution" (Deleuze, 1994, p. 163).

Now, I contend that the same type of argument could be made for the concepts invented in creative mathematics and science. Each mathematical concept (1) emerges from the cooperative confrontation between mathematical thought and chaos; (2) is multi-component; (3) is related to or is a problem; and (4) has a history. Consider a concept of space with its constitutive concept-components, such as point, line, plane, distance (which is necessary to have geometry), and so forth, and the history of the transformation of this concept from Euclid to Descartes (a coordinate space) to Riemann (a space defined as a manifold) to modern topology and geometry. It is sometimes difficult to perceive this multi-component architectural constitution of concepts in mathematics and science, because this complexity can be circumvented in the technical practice of mathematics and science, and in this respect contrasts to philosophy. A more conventional understanding of concepts (such as a generalisation from particulars), when joined with mathematical and scientific propositions, tends to suffice. This may be one of the reasons why Deleuze and Guattari deny the role of concepts – in their sense – in mathematics and science, declaring that "it is pointless to say that there are concepts in science [including mathematics]" (Deleuze & Guattari, 1994, p. 33). For them, mathematical and scientific thought is limited to planes of reference, linked to the invention of functions (or other mathematical entities) and propositions, and lacking planes of immanence, which make concepts possible (p. 132). Although Deleuze and Guattari's view is ultimately more ambivalent than this brief summary suggests, they do not extend their concept of the plane of immanence to mathematics and science. By contrast, I argue that planes of immanence and the creation of concepts (in Deleuze and Guattari's sense) play central roles not only in creative philosophical thought but also in creative mathematical and scientific thought.

I am not disputing that mathematical and scientific thought also works with planes of references, functions, propositions, and so forth. Planes of reference give rise to formations that define the disciplinary nature of mathematics and science, and as such essentially shape mathematical and scientific thinking and practice – essentially but not completely or even most centrally in creative mathematics and science. Rather, I argue that creative thought in *both* philosophy and mathematics or science is defined by planes of immanence and creation of concepts, in Deleuze and Guattari's sense. By the same token, one can maintain the emphasis on *creating* concepts, which carry individual signature underneath them, as much in mathematics or science as in philosophy. Just as there are Descartes' cogito and Leibniz's monads in philosophy, there are Riemann's manifolds or, before him, Descartes' coordinate spaces in mathematics, and Einstein's spaces in relativity, and Heisenberg's matrix variables in quantum mechanics.

Deleuze and Guattari allow that creative mathematical and scientific thought could have philosophical aspects and involve philosophical concepts. But they appear to associate these aspects with *inexact* (but philosophically rigorous) thought within the *exact* (mathematically or scientifically rigorous) thought. This philosophical thought is either operative alongside mathematical and scientific thought or enters by way of "interference" between mathematics and philosophy (pp. 217–218). My argument is different. I argue that creative *exact* mathematical and scientific thought is defined by planes of immanence and invention of (exact) mathematical or scientific concepts, the architecture of which is analogous to that of philosophical concepts in Deleuze and Guattari's sense. *Mathematical* and *scientific* planes of thought are planes of immanence, and the nature of mathematical and scientific concepts is analogous to that of philosophical concepts. In other words, it is not only a matter of mathematical and scientific thought and, thus, concepts becoming philosophical at certain junctures, but a matter of the mathematical and philosophical thought creating parallel homomorphic (partially corresponding to each other), although not isomorphic (fully corresponding to each other), architectures of mathematical and philosophical concepts. These architectures are not isomorphic because technical, exact aspects of mathematical and scientific concepts, demanded by the disciplinary aspects of mathematics and science, are not found in philosophical concepts.

In his rethinking of spatiality and geometry, Riemann introduces a new *mathematical plane of immanence*, a plane that gives rise to multi-component mathematical concepts, along with a *mathematical plane of reference*, a plane that gives rise to frames of references, functions, and propositions. The plane of immanence, as the plane of the movement of thought, is not "a concept that is or can be thought," but is "the image of thought, the image that thought gives itself of what it means to think" (p. 37). As Deleuze and Guattari say, "Concepts are like multiple waves, rising and falling, but the plane of immanence is the single wave that roles them up and unrolls them" (p. 36). The present argument aims to *extend*, rather than to juxtapose, the plane of immanence (and the relationship between it and concepts) to mathematical and scientific thought, and to join this plane with the plane of reference. For Deleuze and Guattari, mathematics or science "relinquishes the infinite, infinite speed, in order to a gain *a reference able to actualize the virtual* [of chaos].... [It] gives reference to the virtual, a reference that actualizes the virtual through functions" (p. 118; translation modified). Thought's enactment of this process constitutes a plane of reference. Planes of reference do play a major role in mathematical and scientific thinking, especially in the disciplinary functioning of mathematics and science. This, however, is not inconsistent with the view that planes of immanence and the creation of concepts are found in mathematics and science. Mathematical and scientific thought combines both planes, that of immanence and that of reference, and creates its concepts from this fusion.[2]

Riemann's thought lays out a new plane of immanence of *mathematical* thought, and this plane extends beyond geometry. Riemann made major transformative contributions of *conceptual* nature (which grounds his major technical contributions) to nearly every area of modern mathematics – geometry, topology, analysis, algebra, and number theory – not the least by bringing these fields to bear on each other. The conceptual character of Riemann's mathematics has been noted and emphasized, especially in contrast to mathematics based in calculations (e.g. Laugwitz, 1999; Ferreirós, 2006). Thus far, however, these commentaries have not paid sufficient attention to the multi-component nature of Riemann's concepts and the plane of immanence of his thinking. I might add that this does not mean that Riemann does not do calculations, which are necessary in mathematics, but only that he gives priority to concepts, even in doing calculations.

THE CONCEPT OF MANIFOLD: SPACE AND GEOMETRY IN RIEMANN

Riemann opens his Habilitation lecture as follows:

> As is well known, geometry presupposes the concept of space, as well as assuming the basic principles for construction in space. It gives only nominal definitions of these things, while their essential specification appears in the form of axioms. The relationships between these presuppositions is left in the dark; we do not see whether, or to what extent, any connection between them is necessary, or a priori whether any connection between them is even possible.
>
> From Euclid to Lagrange … this darkness has been dispelled neither by the mathematicians nor the philosophers who have concerned themselves with it. The reason [ground] [*Grund*] for this is undoubtedly because the general concept of multiply extended magnitudes [*Grösse*], which includes spatial magnitudes, remains completely unexplored. I have therefore first set myself the task of constructing the concept of a multiply extended magnitude from general notions of magnitude. It will be shown that a multiply extended magnitude is susceptible of various metric relations, so that space constitutes only a special case of a triply

[2] As I said, it also involves the plane of composition, found in art, but the subject is beyond my scope. To give one example, a two-dimensional plane is a composition of points, lines, and neighborhoods of points, a fact that Kandinsky artistically explored in his "compositions" (his persistent title).

> extended magnitude. From this, however, it is a necessary consequence that the theorems of geometry cannot be deduced from general notions of magnitude, but that those properties that distinguish space from other conceivable triply extended magnitudes can only be deduced from experience. Thus arises the problem of seeking out the simplest data from which the metric relations of space can be determined, a problem that by its very nature is not completely determined, for there may be several systems of simple data that suffice to determine the metric relations of space; for the present purposes, the most important system is that laid down as a foundation of geometry by Euclid. These data are – like all data – not logically necessary, but only of empirical certainty, they are hypotheses; one can therefore investigate their likelihood, which is certainly very great within the bounds of observation, and afterwards decide on the legitimacy of extending them beyond the bounds of observation, both in the direction of the immeasurably large [*Unmessbargrosse*] and in the direction of the immeasurably small [*Unmessbarkleinen*]. (Riemann, 1854, p. 23; translation modified)

Although brief, these introductory reflections are profound and far-reaching. First of all, Riemann is not interested in the axiomatic approach to geometry and is in fact suspicious of it. As I said, in contrast to most previous work on non-Euclidean geometry, the parallel postulate is not the starting point of his investigation, and is of little significance there. His starting point is that the *concepts* of space and of measuring distances in space (necessary for geometry) have never been adequately defined. This is what leaves "the relationships between [axioms] … in the dark." This questioning was deeper than more customary doubts concerning the parallel postulate, and it was unusual at the time.

The axiomatic approach to geometry is, then, abandoned by Riemann, even though it led to the discovery of non-Euclidean geometry by Lobachevsky and Bolyai. Gauss was unable to establish a possible existence of non-Euclidean geometry because, unlike Lobachevsky and Bolyai, he did not do this for the three-dimensional case. However, the spirit of Gauss' approach to the geometry of two-dimensional surfaces was much closer to Riemann, who was inspired by this approach, which he expressly invoked in the lecture. For Riemann, Gauss was on a right trajectory. In his famous "extraordinary theorem (*theorema egrerium*)," as he called it, Gauss proved that the curvature of a surface was intrinsic to the surface rather than inherited from the ambient space. The theorem suggested that one could see the surface as a curved space in its own right and generalise this concept to higher dimensions. To do so, however, required generalising to higher dimensions the concept of curvature, a key component-concept that defines the geometry of a space, which Riemann was the first to accomplish. This new concept of curvature allows one to endow a given continuous (technically, differential) manifold with both metric (a way of measuring distances) and curvature, which was zero in the case of flat manifolds, corresponding, in the case of two dimensions, to Euclidean space. To define curvature in dimensions three and higher required yet another great invention of Riemann, that being the tensor of curvature, and a new form of differential calculus, tensor calculus on manifolds, which later played a key role in Einstein's general relativity. The independence of the geometry of a given space from its embedding allows one to determine intrinsically whether this space is flat or curved. This is crucial for determining the nature of the physical space in any dimension, extending Gauss' theorem for surfaces, which are two-dimensional manifolds. For example, space is curved in the vicinity of the solar system because of the gravity of the sun and planets, but as I said, it appears to be on average flat at the scale of the observable universe.

Riemann's thinking was, then, conceptual rather than axiomatic: he grounds mathematics, as well as physics, in hypotheses and concepts, such as that of manifolds and its subconcepts (curvature, distance, tensors, and so forth). The problem of space and geometry is now posed in terms of finding concepts that are necessary to define space and to give it geometry. A general concept of manifold is applicable to any possible space, while specific manifolds, flat or curved, define different spaces or subspaces. In this respect, his approach is different from Hilbert's concept of axiomatic "foundations" [*Grundlagen*], an idea that Hilbert also tried to apply to physics (Hilbert, 1900). Riemann's approach arises from a plane of immanence, defined by thinking in terms of concepts, such as and in particular that of manifold, rather than in terms of axioms and propositions, an approach that has defined most thinking concerning geometry before and even after Riemann. Riemann's cluster of manifold-related concepts occupy a plane of immanence in Deleuze and Guattari's sense because they are defined by the movement of thought that gives rise to multicomponent concepts, rather than through creating frames of references, functions, or propositions, as would the plane of reference. As explained earlier, however, in Riemann both planes are ultimately joined, as they must be in the creation of mathematical or scientific concepts. Thus, functions define both the local neighbourhood (as infinitesimally Euclidean) of a manifold and how local spaces in it are connected or pass into each other, as well as, most crucially, metrical relations and curvature, although curvature is ultimately defined by more complex entities – tensors.

While, as I said, the parallel postulate is never mentioned in the lecture, Euclidean geometry is invoked in the lecture as the geometry of "flat space," merely a particular and very special case of geometry, where metrical relations take an especially simple form, defined by the Pythagorean theorem. The metrical relation characterizing a given space defines a possible geometry. Non-Euclidean geometry is introduced as such a possible case of geometry, that of a curved (rather than flat) space of either negative or positive curvature. While the non-Euclidean geometry of negative curvature (hyperbolic geometry) was discovered before Riemann, Riemann's lecture introduced non-Euclidean geometry of positive curvature (elliptical geometry). This was another major mathematical discovery with important physical and cosmological implications, some of which were explored by Einstein.

Riemann defines the concept of space – again, understood by him as a physical space (as against, the concept of manifold, which is mathematical) – in accordance with then available hypotheses, as a three-dimensional instance of the concept of continuous manifold, which I shall now explain. My aim in focusing on this concept is that, while a manifold, as defined by Riemann, may be either discrete or continuous (which are, in effect, different concepts of manifoldness), this concept has a richer and more complex architecture, and most of Riemann's lecture is devoted to it. Technically, Riemann considered differential manifolds, which means that one can define differential calculus in them, and indeed Riemann considered metrical manifolds, now called Riemannian, which allow for the concept of distance between any two points in them and thus for geometry. Riemann's concept of continuous manifolds was a new concept of geometrical multiplicity. It is a multiplicity of local (sub)spaces out of which a given space is composed. A continuous (differential) manifold is understood by Riemann on the model of two-dimensional surfaces, defined by Gauss in terms of their intrinsic geometry. Riemann defined first the concept of "n-dimensional magnitude," which allows one to determine a position in a manifold by n numerical determinations, in the same way a position is determined by coordinates in the Euclidean space of n dimension. A continuous manifold is defined by Riemann as locally, infinitesimally Euclidean. This makes a continuous manifold into a conglomerate of local continuously connected open neighbourhoods around each point. (An open neighbourhood could be conceived based on the model of the two-dimensional circle with its circumference removed.) The concept of neighbourhood (also a technical mathematical term), defined as infinitesimally flat and Euclidean, is a component-concept of the concept of manifoldness.

The idea of composing neighbourhoods into a manifold is naturally extendable to a still more general mathematical concept, that of topological space. In this case local neighbourhoods need no longer be Euclidean or allow for measuring distances between points. In the case of Riemannian or, more generally, differential manifolds, while each neighbourhood is infinitesimally flat, Euclidean, the manifold as a whole, in general, is not, except in the special case of flat, Euclidean manifolds. A manifold may be negatively or positively curved, and – which is another major innovation of Riemann – this curvature can also be variable. Riemann defined the metric form as a quadratic differential function of coordinates, and he assumed that the transition from one local coordinate system to another was differentiable. This is why, as defined by Riemann, continuous manifolds were differential and Riemannian, allowing for measuring distances and thus for geometry.

Although it was Riemann's theory of continuous or, again, differential manifolds that had the greatest impact in geometry and physics, the concept of discrete manifolds was important for Riemann's argument. A discrete manifold has a dimension zero topologically. It may, however, still be seen as multiple extended if it forms a very fine lattice with very small intervals between points, which can be "filled," as it were, to form a continuous space of the corresponding topological dimensions. It is also possible to introduce metrical relations for discrete manifolds, and this is important in considering the nature of physical space or, in Riemann's more precise terms, the physical "reality underlying space." Although Riemann is cryptic on the subject of such metrical relations, he allows that the physical "reality underlying space" might be "a discrete manifold" (Riemann, 1854, p. 33). This possibility has been entertained, sporadically, even before Riemann. Finite geometries were beginning to be developed around Riemann's time as well.

Riemann's contribution to physics in the lecture was nearly as crucial as his contribution to mathematics. According to Riemann, a general concept of manifold, and the concept of metric relations that obtain in a manifold, are established mathematically. These relations define a flat or curved nature of a given manifold, unless a manifold is discrete, in which case the metric relations, which could still be defined, are no longer related to flat or curved manifoldness. Such a concept can then apply to physical space. Riemann considers these applications in the last chapter of the lecture, "Applications to Space." However, he makes clear from the outset that any such application can only be based on hypotheses that we form concerning space, and which we can then experimentally test. These hypotheses ground our thinking concerning space and geometry, although these hypotheses also arise from previous assumptions that we have tested or assumed as self-evident. Hence, the fact "that a multiply extended magnitude is susceptible of various metric relations, so that space constitutes only a special case of a triply extended magnitude" implies that "those properties that distinguish space from other conceivable triply extended magnitudes can only be deduced from experience" (Riemann, 1854, p. 23). By "experience," Riemann means an experimental determination of the nature of the actual space. For example, one may consider whether actual space is Euclidean or not, or in the latter case whether it has a positive or negative curvature, or (a question, again, never posed before Riemann) whether this curvature is permanent or variable.

The crucial part of Riemann's thinking *concerning physics* is that it is essentially based on *hypotheses*, specifically those concerning the character of physical space, or perhaps even the hypothesis that something like space exists – in the first place, that is – physically rather than only phenomenally. Coming in the wake of the discovery of non-Euclidean geometry and from a student of Gauss, Riemann's argument is not that surprising. What was remarkable was Riemann's philosophical boldness in his thinking in, reciprocally, both mathematics and physics. Riemann's assessment of his own project is also of great significance: "Investigations like the one just made here, which begin with general concepts, can *only* serve to insure that this work is not hindered by unduly restricted concepts and that progress in comprehending the connection of things is not obstructed by traditional prejudices" (p. 40; emphasis added). "Only" is hardly necessary here. We do need physics to test and even to form our hypotheses concerning space. But we also need – physics proves that we do! – new concepts to counteract "unduly restricted concepts" and to be able "in comprehending the connection of things" not to be "obstructed by traditional prejudices," such as that of the unlimited validity of Euclid's conception of geometry or Newton's conception of physics. This is how we fight against opinion – always an enemy of thought, as Deleuze and Guattari said. Thus the misfortune of science comes from opinion, too.

CONCLUSION: A PEDAGOGY OF THE CONCEPT

I would like, in closing, to comment, on "a pedagogy of the concept." I cite first Deleuze and Guattari's argument concerning the subject in philosophy:

> Philosophers have not been sufficiently concerned with the nature of the concept as philosophical reality. They have preferred to think of it as a given knowledge or representation that can be explained by the faculties able to form it (abstraction or generalization) or employ it (judgment).

> But the concept is not given, it is created; it is to be created. It is not formed but posits itself in itself – it is self-positing. Creation and self-positing mutually imply each other because what is truly created, from the living being to the work of art, thereby enjoys a self-positing of itself, and an autopoetic characteristic by which it is recognized.... The post-Kantians, and notably Schelling and Hegel, are the philosophers who paid most attention to the concept as philosophical reality in this sense. Hegel powerfully defined the concept by the Figures of its creation and the Moments of its self-positing.... In this way Hegel showed that the concept has nothing whatsoever to do with a general or abstract idea, any more than with an uncreated Wisdom that does not depend on philosophy itself. But he succeeded in doing this at the cost of indeterminate extension of philosophy that, because it reconstituted universals with its own moments and treated the personae of its own creation as no more than ghostly puppets, left scarcely any independent movement of the arts and sciences remaining. The post-Kantians concentrated on a universal *encyclopedia* of the concept that attributed concept creation to a pure subjectivity rather than taking on the more modest task of a *pedagogy* of the concept, which would have to analyze the conditions of creation as factors of always singular moments. If the three ages of the concept are the encyclopedia, pedagogy, and commercial professional training, only the second safeguards us from falling from the heights of the first into the disaster of the third – an absolute disaster of thought whatever its benefits might be, of course, from the viewpoint of universal capitalism. (Deleuze & Guattari, 1994, pp. 11–12)

Riemann was a post-Kantian and, when it comes to concepts, post-Hegelian. Nevertheless, even in philosophical terms, the revolutionary nature of Riemann's concept of manifold, according to Deleuze and Guattari, "marked the end of [Hegelian] dialectic and the beginning of the typology and topology of [manifolds]" (Deleuze & Guattari, 1987, p. 482; translation modified). As such, this concept also opened the way to a pedagogy of concepts, as against the universal encyclopedia of concepts; as such, the task is more modest but also more essential because it is a practical task in mathematics, physics, and philosophy alike. This means (limiting the case to mathematics for the moment) that the task is not to aim at compiling all possible concepts, so one could borrow them from a universal encyclopedia in order to solve problems, but rather to invite one to think, to confront problems with concepts in order to create new concepts. Obviously, this may require a certain repertoire of concepts, but this is not the same as using a universal encyclopedia of concepts, which is in fact not possible. Riemann's thought urges us to change our teaching of geometry, which, virtually without exceptions, still proceeds, at all levels, along axiomatic, Euclidean lines. This change, I would argue, could begin at the earliest stages of mathematical education. We might want to rethink the pedagogical practice in mathematics or mathematical physics as a pedagogy of *concepts*. Deleuze and Guattari's conclusion concerning "the three ages of the concept" is even more pertinent now in relation to mathematics than philosophy. This is because our commercial professional training is governed by mathematics and science, *a certain type* of mathematics and science, and the proliferation and dominance of digital technology is only one, most visible, part of it. This is not an argument against giving a role, even a major role, to mathematics and science, those enabling digital technology included. The task is making this role that of a pedagogy of the concept, as a pedagogy of creative thought, in our never-ending struggle against the dogmatism of opinion, which defines commercial professional training, and tends to dominate our culture. Against this dominance, creative thought must be kept alive, and keeping it alive is the main task of our teaching practice in mathematics and beyond.

REFERENCES

Deleuze, G. (1994). *Difference and repetition*, tr. P. Patton, New York: Columbia University Press.

Deleuze, G. & Guattari, F. (1987). *A thousand plateaus*, tr. B. Massumi, Minneapolis: University of Minnesota Press.

Deleuze, G. & Guattari, F. (1994). *What is philosophy?* tr. H. Tomplinson and G. Burchell, New York: Columbia University Press.

Ferreirós, J. (2006). Riemann's Habilitationsvortrag at the Crossroads of Mathematics, Physics, and Philosophy. In J. Ferreirós and J. J. Gray (Eds), *The architecture of modern mathematics: Essays in history and philosophy* (pp. 67–96). Oxford: Oxford University Press.

Hilbert, D (1900). *Foundations of geometry*, 10th edition, tr. Leo Unger. La Salle, IL: Open Court.

Laugwitz, D. (1999). *Bernhard Riemann: Turnings points in the conception of mathematics*, tr. A. Shenitzer. Boston, MA: Birkhäuser.

Riemann, B. (1854). On the Hypotheses That Lie at the Foundations of Geometry. In P. Pesic (Ed.), *Beyond geometry: Classic papers from Riemann to Einstein* (pp. 23–40). Mineola, NY: Dover, 2007.

6

Deleuze and the Conceptualisable Character of Mathematical Theories

SIMON B. DUFFY

In this chapter I present an account of what a mathematical concept is, as developed in the work of the French philosopher Gilles Deleuze. In characterising his understanding of what a mathematical concept is, I draw upon the way that he taps into the mathematical philosophy of Albert Lautman and one of Lautman's contemporaries, Jean Cavaillès. To make sense of what Deleuze understands by a mathematical concept requires unpacking what he considers to be the conceptualisable character of a mathematical theory. However, his account of mathematical theories, indeed his account of mathematics, is much broader than what is usually understood by these terms. Mathematics is more than just the sum of its theories, and mathematical theories are determined by more than just the proofs upon which the theories are based. Deleuze identifies a problem-solution complex in operation at the base of every theory.

By problem-solution complex Deleuze does not mean a 'problem set' and the written 'solutions' to them that are encountered in pedagogical settings, but rather the mathematical problems that emerge during, and indeed characterize, the historical development of mathematics, to which mathematicians have responded by developing more or less rigorous solutions in the form of mathematical concepts derived from: conjectures, for which proofs have yet to be developed; proof, either of conjectures or of other mathematical problems in general; formal theories, based on the proofs; and informal, or semi-formal, theories, for which no rigorous proof has developed, or at least not for all aspects of the proof. Informal theories rely rather on heuristic proofs or inductive approximations derived intuitively. For Deleuze, the mathematical problems, for which a theory is a solution, retain their relevance to the theories not only as the conditions that govern their development but also insofar as they can contribute to determining the conceptualisable character of those theories. Deleuze presents two examples of mathematical problems that operate in this way. These two are the problem of the solvability of quintics and the problem of the diagrammatic representation of essential singularities.

The first is formal: Galois' group-theoretic proof of the solvability of higher-degree polynomial equations. Galois proved that, despite some specific quintics actually having solutions, it was impossible to construct an algebraic formula that solved all quintics. And the second is informal: Poincaré's qualitative theory of differential equations which develops the concept of an essential singularity. The different kinds of essential singularity are determined by virtue of the observed trajectories of variables across a potential function, rather than by a formal mathematical proof. The diagrammatic representation of essential singularities, which are determined in relation to the problem of the representation of meromorphic functions, remains only intuitive. The question of the solvability of quintics and the diagrammatic representation of essential singularities are each examples of mathematical problems, the conditions of solvability of which are immanent to the problems themselves. Deleuze considers these mathematical problems, and the mathematical concepts derived from them, to be characteristic of a more general theory of mathematical problems.

This distinction between formal and informal characterizations of the mathematical expression of problems is important for determining how Deleuze's approach to the relation between mathematics and philosophy differs from those committed to foundationalist approaches to the philosophy of mathematics. Deleuze traces the development of what he characterizes as a more general theory of mathematical problems through the history of mathematics drawing upon the work of Karl Weierstrass, Henri Poincaré, Bernhard Riemann, and Hermann Weyl, and the historical insights of Lautman and Cavaillès. An account of what a mathematical concept is for Deleuze will be developed by tracing the contours of this history and of its broader significance to his work, chief amongst which is showing how such a theory of mathematical problems can be deployed as a way of studying problems in other discourses, or fields and contexts.

Deleuze develops an account of mathematical theories, and thus of the concepts derived from them, that presents them as responses to problems, the immanent conditions of which not only govern the very development of the concepts but continue to be implicated in the concepts as the conditions of their development (Deleuze, 1994, p. 160). The implications are that mathematical concepts are not solely derived from theories that resolve mathematical problems (Deleuze, 1994, p. 158), but rather mathematical concepts are themselves *problematic* insofar as the mathematical problems that govern the development of the theories continue to be implicated in the concepts, as the conditions of their very development. Deleuze wants to challenge any presumption that mathematical problems disappear as problems, or can be abandoned, with the discovery of their solution, and therefore also what this presumption falsely entails, that it is the truth of the solution that retrospectively governs the development of the problems.

While Deleuze would acknowledge that most mathematical concepts do seem to fit with this model, i.e. of dissolving the problem and retrospectively governing the development of the problem for which they are the solution, he would maintain that this seeming fit is misleading and at best a convenient shorthand. It is misleading insofar as it sets up the expectation that all problems will have solutions, or rather that all adequately determined problems will have solutions. When the expectation of a solution is thwarted, the fault is considered to lie with the particular characterisation of the problem, rather than with the expectation of a solution. Deleuze considers this to be a poor model for a theory of mathematical problems, as it fails to account for those problems that are able to be characterised adequately, i.e. with generally acceptable mathematical rigour, and yet whose expectation of solution is thwarted, i.e. the associated mathematical proof disproves a solution or proves that there is no solution. The example that Deleuze uses is Galois' group-theoretic proof of the solvability of higher-degree polynomial equations. The conceptualisable character that Deleuze extracts from these mathematical concepts is presented as a theory of mathematical problems, and would be something like this: Well-defined problems don't necessarily have rigorous solutions, indeed informal diagrammatic representations of results do not provide solutions, but rather characterise the conditions of the problem itself. Deleuze argues that these kinds of concepts are responses to the particular conditions of the problem itself and are governed by those conditions rather than by an actual solution whose discovery is expected, since there is no such solution. Deleuze considers this example to function as an effective counterexample to what I have characterised as the poor model, and wants to generalise from this particular case to a general theory of mathematical problems.

Deleuze proposes the account of mathematical concepts that he develops as a more inclusive alternative. The implication is that mathematical concepts are not solely derived from theories that are the expected solutions to mathematical problems, but rather mathematical concepts are themselves problematic insofar as the mathematical problems that govern the development of the theories from which the mathematical concepts are derived continue to be implicated in the concepts as the conditions of their very development. So problems are not forgettable, or able to be discarded or abandoned, relative to their solutions. Nor are they simply useful for pedagogical reasons, i.e. to educate budding mathematicians about solutions, which are then discarded or seamlessly displaced by the solution itself. Mathematical problems are rather intrinsic features of the enduring problem-solution complex.

Deleuze then uses this theory of mathematical problems as a model to develop a general theory of problems, and specifically philosophical problems, with the aid of tracing an alternative lineage in the history of philosophy that tracks these developments in mathematics. For example, (1) Maimon's account of intuitions as differentials allows Deleuze to incorporate a critique of representation within the structure of his philosophy, to which I return at the end of this chapter; and (2) Riemann's work allows Deleuze to critique and reconfigure Kant's pure intuitions of space and time along Bergsonian lines, although arguably modelled more effectively on the mathematics that inspired Bergson.[1]

THE THEORY OF MATHEMATICAL PROBLEMS AND THE HISTORY OF MATHEMATICS

Deleuze traces the development of his alternative account of a theory of mathematical problems through the history of mathematics drawing upon the work of Weierstrass, Poincaré, Riemann, and Weyl. This history actually begins with the Leibnizian method of approximation (1684) using successive orders of the differential relation, which was developed into a theorem about power series expansions by Brooke Taylor in 1715, for which Lagrange attempted to provide an algebraic proof in 1772 (1797). "Lagrange tried to make the calculus rigorous by reducing it to the algebra of infinite series" (Grabiner, 1983, p. 203). The foundation of algebra was generally thought to be sound in the eighteenth century. Lagrange thought he had proved that every differentiable function was the sum of a Taylor series. He coined the expression 'derived function' and the notation for it f'(x), and recursively f"(x), etc., which is the origin of the term 'derivative.' The concept of function helped 'free the concept of derivative from the earlier ill-defined notions' – Leibniz's differential quotient, a ratio with vanishing quantities; and, Newton's fluxion – since the derivative, as a function, was "the same sort of object as the original function" (Grabiner, 1983, p. 203).

[1] See Duffy, 2013, pp. 89–116.

However, the problem with Lagrange's method was that it assumed that "the algebra of finite quantities' could 'automatically be extended to infinite processes" (Grabiner, 1983, p. 203), which turned out not to be the case. The subsequent development of the epsilon-delta method by Weierstrass, which reformulated the calculus without either fluxions or vanishing quantities using only real numbers, allows Lagrange's method to be formalised in the calculus by Weierstrass' theory of analytic continuity (1872), which is a theory of integration as the approximation of functions from differential relations according to a process of summation in the form of series, using Taylor series or power series expansions.

The history that Deleuze traces incorporates Poincaré's qualitative theory of differential equations (1881–1882), which provides a diagrammatic response to problems with the representation of meromorphic functions, or divergent series, in Weierstrassian analytic continuity, and further extends to Riemann's concept of qualitative multiplicity, or Riemann space (1854). The work of Weyl on Riemann surfaces (1913)[2] is instrumental to the development of the mathematical model that Deleuze develops.[3] Weyl makes Riemann's intuitive representation of Riemann space more explicit by using a generalisation of Weierstrass' analytic continuity – effectively demonstrating that Riemann surfaces are the surfaces of Weierstrassian power series expansions – to show that Riemann space is composed of Riemann surfaces, and therefore of Weierstrassian power series expansions.[4] It is by tracing the history of the philosophical engagements with and responses to these mathematical developments, and by drawing upon the work of Lautman and Cavaillès, that Deleuze develops a theory of mathematical problems, which he then uses as a model for a theory of problems in general as they arise in other discourses.

It is important to note that Deleuze eschews characterising his engagement with mathematics as simply analogical or metaphorical. He is careful to distinguish between those mathematical concepts that are quantitative and exact in nature, which he considers it to be 'quite wrong' to use metaphorically 'because they belong to exact science' (Deleuze, 1995, p. 29), and those mathematical problems that are 'essentially inexact yet completely rigorous' (Deleuze, 1995, p. 29) and which have led to important

[2] See Weyl, 1964.

[3] Deleuze is aware of Weyl's work on Riemann via Lautman's commentary on Weyl, which Deleuze cites in Deleuze and Guattari, 1987, pp. 485. See Lautman, 2011, pp. 133–137 and Duffy, 2013, pp. 103–115. The importance of Lautman's work to Deleuze's engagement with mathematics in Deleuze, 1994 is explored in Duffy, 2013, pp. 117–136.

[4] Deleuze extends this model with Poincaré's qualitative theory of differential equations to characterise the relations between discontinuous (Riemann) surfaces of Riemann space.

developments not only in mathematics and science in general but also in other nonscientific areas such as philosophy and the arts. Deleuze argues that this sort of concept is "not unspecific because something's missing but because of its nature and content" (Deleuze, 1995, p. 29). An example of an inexact and yet rigorous concept is Poincaré's qualitative theory of differential equations which develops the concept of an essential singularity. The different kinds of essential singularity are determined by virtue of the observed trajectories of variables across a potential function, rather than by a specific formal mathematical proof. Another example is a Riemann Space. Bernhard Riemann generalises Gauss' work on the differential geometry of surfaces – namely, that the curvature of a surface embedded in three-dimensional space may be understood intrinsically to that surface, i.e. independently of the three dimensional space in which it is embedded – into higher dimensions. While Euclidean 'finite' geometry holds for three-dimensional linear point-configurations, curved three-dimensional spaces are not necessarily flat. However, these spaces still resemble Euclidean space in the infinitesimal neighbourhood of each point. By considering the infinitesimal neighbourhood around each point as a small bit of Euclidean space, the entire space can then be constructed with the step-by-step juxtaposition, or accumulation, of these infinitesimal neighbourhoods. The resulting Riemannian space can be defined as an assemblage of local spaces, each of which can be mapped onto a flat Euclidean space, without this determining the structure of the manifold or multiplicity as a whole.

While Deleuze recognises that citing mathematical concepts of the exact kind outside of their particular sphere would rightly expose one to the criticism of "arbitrary metaphor or of forced application" (Deleuze, 1989, p. 129), he defends the use he makes of mathematical concepts of the inexact kind. He does so on the grounds that by taking from these mathematical concepts "a particular conceptualizable character which itself refers to non-scientific areas" (Deleuze, 1989, p. 129), the redeployment of this conceptualisable character in a nonscientific area is justified. What this means is that the other nonscientific area "converges with science without applying it or making it a metaphor" (Deleuze, 1989, p. 129). A useful way of characterising the relation between the conceptualisable character of the inexact mathematical concept and this conceptualisable character as redeployed in other nonscientific areas, insofar as the latter converges with the former, is to refer to it as a modelling relation. That is, the conceptualisable character which is redeployed in a nonscientific area is modelled on the conceptualisable character of the inexact mathematical concept. What distinguishes a modelling relation from a relation of analogy or metaphor is that there are "correspondences without resemblance" (Deleuze, 1994, p. 184) between them. That is, there is a correspondence between the conceptualisable character in each instance; however, there is no *resemblance* between the mathematical elements of the mathematical problem and the non-mathematical elements of the discourse in which this conceptualisable character has been redeployed. It is the *conceptualisable character* of the two examples above, or at least of how the former is implicated in the latter, that Deleuze redeploys in his philosophy. He is interested in the conditions of the discontinuity between the two discontinuous analytic functions in Poincaré's qualitative theory of differential equations, which, post Weyl, would characterise the relations between discontinuous (Riemann) surfaces of a Riemann space.

While Deleuze does refer to his project as developing a "*mathesis universalis*" (Deleuze, 1994, p. 181), he does not consider there to be a definite system of mathematical laws at the base of nature. Mathematics is not privileged in this way over other discourses. There is, however, a peculiarity about the discourse of mathematics that distinguishes it from other discourses, and that is the very general nature of the relation between the objects of the discourse and the ideas that we have of those objects as expressed within the discourse. Mathematics is peculiar insofar as all of its objects are actually constructed by the discourse itself. By this I just mean very generally that they are the product of the discursive practice of mathematics by mathematicians; they are not discovered empirically. I take it as uncontroversial that there are mathematical objects, and that these objects are abstract. The ideas that we have of the objects of mathematics are therefore directly and unproblematically related to the objects themselves. And this is regardless of subsequent questions about the status of those objects, from the point of view of the philosophy of mathematics: questions about the independence of those objects, whether we are talking about objects or structures, or even about competing constructions in mathematics itself. It is for this reason that mathematics is figured as providing a model for our understanding of the nature of this relation, between the objects of a discourse and the ideas of those objects as expressed within the discourse, in discourses other than mathematics, where this relation is far from straightforward.

LAUTMAN, CAVAILLÈS AND THE MATHEMATICAL REAL

What is important about mathematics, for Lautman and Deleuze, is its seeming *a priority*, which allows the structure of problematic ideas, or the theory of mathematical problems, to be recognised as a component of the mathematical real in a way that is not directly accessible in other discourses because of the reasons mentioned in the paragraph above. Deleuze takes Lautman's concept of the mathematical real, which includes the sum of all mathematical theories and the structure of the problematic ideas that govern them, as the basis for his reflections on the theory of problems, and casts it as a model for our understanding of the nature of the relation between the objects of any one discourse and the structure of the problematic ideas that govern them within that discourse. Insofar as he claims that all discourses can be modelled in this way, Deleuze argues that there is a "*mathesis universalis*" (Deleuze, 1994, p. 181). Deleuze is not positing a positive mathematical order to the universe, but he is rather nominating the Lautmanian mathematical real, and the theory of problems that he characterises by means of it, as a model for our understanding of the structure of other discourses.

What does Deleuze mean by the structure of the problematic ideas that govern the development of mathematical theories, which is included in Lautman's concept of the mathematical real and, that I argue, forms the basis of Deleuze's theory of problems? Lautman subscribed to a Platonic understanding of the structure of the problematic idea; however, he understands these to be abstract dialectical ideas, i.e. not universal Forms, but Archetypes or Ideals, which are the touchstones for the selective and organizational function of the Dialectic,[5] and which remain revisable in the face of the demands of that organization. The abstract dialectical ideas to which Lautman is referring include the following: local-global; intrinsic-extrinsic; essence-existence; continuous-discontinuous; and finite-infinite.

It is important to note that Lautman's references to a dialectic of ideas should not be understood as being references to a general dialectic that exists independently of the mathematics. Lautman is quite explicit in claiming that the dialectic of ideas is the fourth point of view of the mathematical real (Lautman, 2011, p. 183). For Lautman, Ideas constitute, along with mathematical facts, objects and theories, a fourth point of view of the mathematical real. "Far from being opposed these four conceptions fit naturally together: the facts consist in the discovery of new entities, these entities are organized in theories, and the movement of these theories incarnates the schema of connections of certain Ideas" (Lautman, 2011, p. 183). For this reason, the mathematical real depends not only on the base of mathematical facts but also on dialectical ideas that govern the mathematical theories

[5] According to Lautman, they are "the structural schemas according to which effective theories are organized" (Lautman, 2011, p. 199).

in which they are actualized. The mathematical real is not just the sum of all mathematical theories. The former should therefore not be collapsed into the latter. To do so would lead to the mistaken thesis that mathematics provides evidence of an external and more general dialectic that is equally accessible by means of some kind of analysis performed in regard to or from within other discourses. What seems to be clear in Lautman's work is that he considers himself to be working within the constraints of the discourse of mathematics, and the structure of the dialectic that he presents is determined as operating within the expanded concept of mathematics that he makes claim to: the mathematical real. The dialectic of ideas is independent of the mathematical theories, or the mathematics per se, but not of the expanded understanding of the mathematical real.

Lautman does claim that the structure of the dialectic is not the sole privy of the mathematical real, and that it can therefore also 'be found' in other discourses. However, he does not claim that this is the case because the dialectic is able to be generalised, or insofar as it is transcendent with respect to the mathematical real. While Lautman makes strong claims to the unity of mathematics, which was controversial at the time and remains so today, he does not make any claim whatsoever as to the unity of all discourses. What Lautman argues rather is that this is the case because of the way the structure of the dialectic operates in the mathematical real functions as a model for recognizing how it can be understood to operate in other discourses.

Lautman maintains that we are able to recognise the logic of relations structured by the dialectic in other discourses solely by virtue of the mathematical theories in which these relations are incarnated; as he argues, "the effectuation of these connections is immediately mathematical theory" (Lautman, 2011, p. 28). That is to say that it is the way in which the mathematical logic is deployed in other discourses that allows such a discourse to be understood to operate according to the dialectic. This is perhaps straightforward in those discourses that deal extensively with mathematics, such as the sciences and social sciences, even philosophy; but not so straighforward in those that do not, such as those under the umbrella of the humanities. But even a simple concept like representation, or the problem of the relation between an object and the idea that we have of that object, referred to above, can be understood to be modelled on mathematics in the requisite way being referred to here by Lautman, and therefore operate according to the dialectic. A more thorough treatment of the problem of representation will be returned to at the end of the chapter. By dialectic Lautman means here the dialectic of the mathematical real. So mathematics is not privileged over other discourses according to Lautman because, on the one hand, he does not consider there to be a definite system of mathematical laws at the base of nature, and, on the other hand, he does consider it to be intimately involved in our understanding of the very dialectical structure of those discourses. What this amounts to is that mathematical theories are not the sole privy of mathematics, or the mathematical real; they also provide the ground for understanding how the dialectic operates in other discourses. So when Lautman argues that "mathematical logic does not enjoy in this respect any special privilege. It is only one theory among others and the problems that it raises or that it solves are found almost identically elsewhere" (Lautman, 2011, p. 28) – by privileged we should also understand exclusive to the mathematical real.

While Lautman subscribed to a Platonic understanding of the structure of the problematic idea, Deleuze also draws upon the comments of Cavaillès, who is critical of this aspect of Lautman's work, to characterise problematic ideas as being immanent to the problems themselves.[6] The example from mathematics that Deleuze uses to characterise the immanent nature of problematic ideas, and thus of a more general theory of mathematical problems, is the question of the solvability of polynomial equations.

GALOIS (FORMAL) OR POINCARÉ (INFORMAL) ON THE MATHEMATICAL EXPRESSION OF PROBLEMS?

Lagrange provided a unified understanding of the general formulas for determining the solutions to polynomial equations of degree less than or equal to four. However, Lagrange was unable to do the same for quintics, and suggested that they might not be solvable in this way. Abel provided the first conclusive proof of this conjecture. He proved that, despite some specific quintics actually having solutions, it was impossible to construct an algebraic formula that solved all quintics. So, the question of the solvability of quintics provides an example of a mathematical problem, the conditions of which are immanent to the problem itself, and which is characteristic of a more general theory of mathematical problems.

Galois developed a more complete theory on the solvability of higher-degree polynomial equations, by showing that a simpler proof of Abel's

[6] 'Cavaillès: "Personally I am reluctant to posit something else that would govern the actual thinking of the mathematician, I see the exigency in the problems themselves. Perhaps this is what he calls the Dialectic that governs; if not I think that, by this Dialectic, one would only arrive at very general relations ... The future will show which of us is right." (Lautman, 2011, p. 224).

result could be found along purely group-theoretic lines. Deleuze argues that Galois' theory is not simply another example or expression of the theory of mathematical problems, but rather a formal restatement of the theory of problems as such, in purely group-theoretic terms. Galois' theory can be understood to unite in one formal theory those aspects of the theory of problems that Deleuze draws upon in Poincaré's qualitative theory of differential equations. Galois' theory allows the formal presentation of a feature that remains only intuitive in Poincaré's qualitative theory of differential equations, namely the leap of the variable across the cut of the potential function in the diagrammatic representation of essential singularities, which are determined in relation to the problem of the representation of meromorphic functions, where more formal solutions remain elusive.

With Galois we obtain a formal proof of a theory of mathematical problems. However, the differences in kind between differential calculus and group theory, namely the formalisation of the latter versus the informal intuitive results of the former, are 'merely secondary' in Deleuze's reckoning. What is important is that they are each characterisations of the mathematical expression of problems as such. So even though Galois' theory is a formal restatement of the theory of problems in purely group-theoretic terms, throughout his work Deleuze prefers and continues to draw upon the informal model that brings together Weierstrass, Poincaré and Riemann, and remains quiet, or is epistemically modest, in regards to mathematical foundations.

This sets up the mathematical real, and the structure of the theory of problems as it operates in the mathematical real, to function as a model for the structure of other discourses, and for how we can understand these other discourses to operate. It is the conceptualisable character of the theory of mathematical problems as a theory of problems that allows other discourses to be understood by the non-mathematician to operate according to a theory of problems, or to be structured by a theory of problems modelled on the theory of problems in mathematics.

What distinguishes Deleuze from those committed to foundational approaches to the philosophy of mathematics is that Deleuze's claims are not dependent on the success of one or other of the foundational approaches to provide foundations for mathematics. Deleuze's claims are rather epistemically modest. Rather than being drawn into making the claim that the model he is proposing is *the* mathematical model, which is defendable by virtue of it being underpinned by a particular foundational approach to mathematics, for Deleuze, the proposed informal model is simply *a* mathematical model that he considers to be more useful than other potential models.

The work of Deleuze does more than merely provide a descriptive account of a theory of problems as the foundations for our understanding of the operations of all discourses. Rather, his work develops an argument for a particular kind of theory of problems that can be understood to operate in relation to other discourses by virtue of the way that it operates as a theory of problems in mathematics, here borrowing the expanded account of mathematics derived from Lautman's concept of the mathematical real. The detail of the structure of the theory of problems can only be offered in mathematical terms; however, the structure of this theory of problems can be used to model the structure and mode of operation of problems in other discourses.

For Deleuze, the way that a mathematical theory, and the mathematical concept derived from it, is implicated immanently in the conditions of the problem that determines it serves as a model for the way that a philosophical concept is implicated in the philosophical conditions of the problem which determines it. There is, therefore, a correspondence between the structure of the theory of problems in mathematics and the structure of the theory of problems that is deployed in Deleuze's philosophy, insofar as the latter is modelled on the former. There are "correspondences without resemblance" (Deleuze, 1994, p. 184) between them, insofar as the structure of the theory of problems conditions each discourse, but without there being a resemblance between the respective problems that do the conditioning.

The philosophical implications of this modelling relation are developed by Deleuze[7] in his critique of representation, or of the presumption of a connection between an idea and something in the world that it represents, presented in *Difference and Repetition*. For Deleuze, the 'idea' is not bound to the representation of an object or a concept, nor is it the property of individual consciousness. Difference, in Deleuze's sense of the term, is also not tied to representation, thus it does not involve a comparison of one thing or concept to another. Deleuze insists that "[d]ifference is not and cannot be thought in itself, so long as it is subject to the requirements of representation" (Deleuze, 1994, p. 262). One aspect of the way that mathematics models the theory of problems for Deleuze is that the mathematics that he draws upon to develop this model also

[7] In *Expressionism in Philosophy* (Deleuze, 1990) in relation to his reading of Spinoza's theory of relations in the *Ethics* (see Duffy, 2004; 2006), and in *Bergsonism* (Deleuze, 1991), and *Cinema 1* and *2* (Deleuze, 1986; 1989) in relation to his understanding of Bergson's intention "to give multiplicities the metaphysics which their scientific treatment demands" (Deleuze, 1991, p. 112). But I want to focus here on how this modelling relation operates in *Difference and Repetition* (Deleuze, 1994).

actually models the nature of the illusory relation of representation between an idea and that which it represents in discourses other than mathematics. Deleuze draws upon the work of Salomon Maimon on the concept of the differential to develop this aspect of his critique of representation. According to Maimon, the operation of integration functions as a mathematical rule of the understanding that is applied to the elements of sensation, which are modelled on differentials, in order to account for how the manifolds of sensation are brought to consciousness as sensible objects of intuition. Here, the determinate units of different manifolds of sensation are projected to be qualitatively different differentials. What appear to us as external objects are therefore constructed as such by the understanding, and the retrospective explanation of the construction is that it is the result of the application of a mathematical rule of the understanding to the elements of sensation.

In the first step of the process, two different manifolds of sensation characterised by different differentials are brought into consciousness by virtue of the application of integration as a rule of the understanding to the elements of sensation or differentials. The method of integration that Maimon deploys as a rule of the understanding to proceed from differentials to functions, or from projected elements of sensation to the qualities of sensible objects, is the method of approximation of a differentiable function around a given point provided by the process of summation in the form of a Taylor series or power series expansion, what is eventually formalised by Weierstrass as analytic continuity. Deleuze champions the inexact extension of this method in Poincaré's qualitative theory of differential equations. The real relation between the two qualities themselves, as sensible objects, is modelled on the mathematical relation between their differentials. A primary physical judgement is then made about the products of integration which determines them as sensible objects. What this amounts to is that all physical judgements, whatsoever, are predicated on a prior mathematical judgement, which "escapes consciousness" (Gueroult, 1929, p. 64). It is therefore an illusion that sensible or real objects appear as external objects to us, when in fact they are the product of our understanding.

What can be seen in operation here is an example of both how a mathematical theory is deployed in another discourse and of the conceptualisable character of the mathematical theory. In other words, Poincaré's qualitative theory of differential equations, which is that a function can be generated from differentials, corresponds structurally to the problem as articulated in another discourse, which is that objects can be generated from the elements of sensation, without there being any resemblance between the respective problems that do the conditioning, and therefore how such a discourse, and indeed all discourses relying on such a form of representation, can be understood to operate according to the dialectic. It should be clear that the conceptualisable component which is deployed in the non-mathematical discourse is modelled on the conceptualisable component of the inexact mathematical concept.

In summary, a mathematical concept for Deleuze is the conceptualisable character of a mathematical theory. He develops an account of mathematical theories, and thus of the concepts derived from them, that presents them as responses to problems, the immanent conditions of which govern not only the very development of the theories, but continue to be implicated in the theories as the conditions of their development. The implication being that mathematical concepts are not solely derived from theoretical resolutions of mathematical problems, but rather mathematical concepts are themselves problematic insofar as the mathematical problems that govern the development of the theories, from which the mathematical concepts are derived, continue to be implicated in the theories as the conditions of their very development. It is on the basis of this account of mathematical concepts as problematic that Deleuze considers the mathematical problems that he identifies in the developments of Galois, Weierstrass, Poincaré, Riemann, and Weyl, and the mathematical concepts derived from them, to be characteristic of a more general theory of mathematical problems. By tracing the history of the philosophical engagements with and responses to these mathematical developments, and by drawing upon the work of Lautman and Cavaillès, Deleuze develops a theory of mathematical problems, which he then uses as a model for a theory of problems in general as they arise in other discourses.

Deleuze does consider the theory of mathematical problems to have been formalised in Galois' group-theoretic proof of the solvability of higher-degree polynomial equations. However, the mathematical problem of primary interest to Deleuze, and which features prominantly in the model he constructs, is not ultimately reducible to a formal proof. The theory of problems derived from Deleuze's informal model is the same as that furnished by a group-theoretic approach, it just arrives at this theory by other means, albeit informal mathematical means. What Deleuze gains by this selection is a mathematical model that affords him a critical perspective on representation not available in the group-theoretic approach.[8] Deleuze

[8] This is not to rule out that other possible critical perspectives on representation cannot be drawn from the group-theoretic approach – just that Deleuze opts for a model furnished by the informal approach.

can be understood to find in the informal mathematical problem that he selects a way to use the relevance of sense experience in obtaining informal results for mathematical problems, to model the operation of sense experience relevant not only to science but to all discursive practices relying on representation.

REFERENCES

Burgess, J. (2014). Quine's philosophy of logic and mathematics. In G. Harman & E. LePore (Eds.), *A companion to W.V.O. Quine*. London: Wiley-Blackwell.

(1986). *Cinema 1: The movement-image* (H. Tomlinson & B. Habberjam, Trans.). London: Athlone. Original work published 1983.

(1989). *Cinema 2: The time-image* (H. Tomlinson & R. Galeta, Trans.). London: Athlone. Original work published 1985.

(1990). *The logic of sense* (M. Lester with C. Stivale, Trans., C. Boundas, Ed.). New York: Columbia University Press. Original work published 1969.

(1990). *Expressionism in philosophy: Spinoza* (M. Joughin, Trans.). New York: Zone. Original work published 1968.

(1991). *Bergsonism* (H. Tomlinson& B. Habberjam, Trans.). New York: Zone. Original work published 1966.

(1994). *Difference and repetition* (P. Patton, Trans.). London: Athlone. Original work published 1968.

(1995). *Negotiations, 1972–1990* (M. Joughin, Trans.). New York: Columbia University Press. Original work published 1990.

Deleuze, G. & Guattari, F. (1987). *A thousand plateaus: Capitalism and schizophrenia* (B. Massumi, Trans.). Minneapolis: University of Minnesota Press. Original work published 1980.

Duffy, S. B. (2004). Schizo-math: The logic of differen*t*/ciation and the philosophy of difference. *Angelaki*, 9(3), 199–215.

(2006). The mathematics of Deleuze's differential logic and metaphysics. In Duffy (Ed.), *Virtual mathematics: The logic of difference* (pp. 146–178). Manchester: Clinamen.

(2013). *Deleuze and the history of mathematics*. London: Bloomsbury.

Grabiner, J. V. (1983). The changing concept of change: The derivative from fermat to Weierstrass. *Mathematics Magazine*, 56(4), 195–206.

Gueroult, M. (1929). *La philosophie transcendantale de Salomon Maimon*. Paris: Alcan.

Lautman, A. (2011). *Mathematics, ideas and the physical real* (S. B. Duffy, Trans.). London: Continuum. Original work published 2006.

Riemann, B. (1963). On the hypotheses which lie at the bases of geometry (W. Clifford, Trans.). *Nature*, 8(183–184), 14–17, 36–37.

Weyl, H. (1964). *The Concept of a Riemann surface* (G. MacLane, Trans.). Reading, MA: Addison–Wesley. Original work published 1955.

PART IV

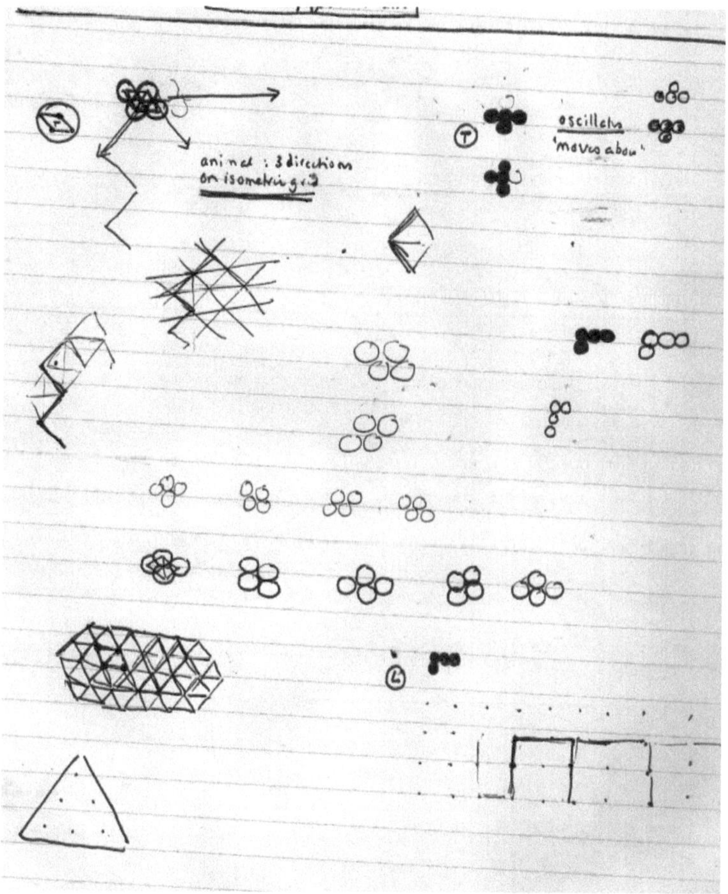

animal: 3 directions
on isometric grid
oscillators
'moves about'

7

Homotopy Type Theory and the Vertical Unity of Concepts in Mathematics

DAVID CORFIELD

INTRODUCTION

Studying the articulation of mathematical concepts is an important task for philosophy, not least for making us lift our eyes above the level of proof to consider the larger enterprise of doing mathematics. Where so much of the Anglophone philosophical literature has dwelt on mathematical activity as essentially a matter of proof construction, the larger top-down view of mathematics as a tradition of intellectual enquiry, rooted many millennia in the past, is typically neglected. Even Lakatos, that pioneer of the study of conceptual change, took proof construction and criticism to be the central motors of conceptual development, rather than tackle larger strategic units of research. My earliest published works in philosophy (Corfield, 1997, 1998) aimed at showing the limitations of this outlook by posing and attempting to answer the question of how we might understand larger-scale activity, such as the mathematical equivalents of the 'research programmes' Lakatos himself had studied in the field of natural science.

By the time I came to write *The Importance of Mathematical Conceptualisation* (Corfield, 2001) it was with a view to direct attention towards yet higher levels of research activity. My examination of 'mathematical research programmes' had led me to see the enormously intricate patterns of lines of influence between mathematics generated in different fields. I chose then to take up as a case study forms of persuasive argument for the value of the concept of 'groupoid', since there were multiple lines of support for this concept from clusters of researchers working in different fields. There was also a countervailing current aiming to deny this value, which maintained that groupoids added little or nothing not already to be found in the concept of 'group'. Here then was to be found a debate not as to the correctness of pieces of mathematical reasoning, but one which concerned the worth of a certain extension of an existing concept. Groupoids, it was claimed by their advocates, had certain key advantages over groups when it came to capturing aspects of symmetry.

In the years since I wrote that piece, groupoids and their higher cousins have been proposed as the building blocks of a new foundational programme for mathematics, known as *homotopy type theory*, one which sees sets demoted from their dominant position to become a lowly component of a larger picture. If things work out well for this programme, it will be a great vindication of the claims made by those who glimpsed the potency of the groupoid concept, even if they never quite envisaged this lofty new role.

In this chapter I want to develop two theses:

1. This new foundational language, homotopy type theory, provides an important perspective from which to understand varieties of mathematical concept.
2. Mathematics displays *vertical unity*, so that concepts met in elementary and in research level mathematics are related.

Thesis 1 requires enough prefatory remarks on what is intended here by 'varieties of mathematical concept' to warrant its own section, but I can quickly sketch now what is meant by Thesis 2. *Vertical unity* is a term introduced by the mathematician Alexander Borovik to indicate that while the continued unfolding of mathematics at the research front appears to give rise to fiendishly complicated ideas, the products of this research in fact never completely depart from the kinds of concepts accessible to those with little mathematical training. This unity, Borovik claims, means that "'recreational', 'elementary', 'undergraduate' and 'research' mathematics are no more than artificial subdivisions of a single continuous spectrum" (Borovik, 2005, p. 1). Where we may take justifiable delight in the 'horizontal' unity provided by far-flung analogies between apparently different fields,

> [t]he vertical unity of mathematics, with many simple ideas and tricks working both at the most elementary and at rather sophisticated levels, is not so frequently discussed – although it appears to be highly relevant to the very essence of mathematics education. (Borovik, 2005, p. 10)

I cannot speak for any other than the philosophical literature in this regard, but already this lack of attention there is interesting. I suspect that the standard representation of mathematics through the filter of first-order logic and set theory is to blame. By the time one has rendered conceptually similar pieces of mathematics into set theory, little or no sign remains of their resemblance. We will see then that Thesis 1 is highly relevant to Thesis 2, since homotopy type theory offers a more faithful representation of mathematics, enough to allow commonality of structure to show through.

After the next section on varieties of concept, I first illustrate this vertical unity with a case which takes its starting point from a commonality found between elementary logic and elementary arithmetic. It transpires that the common structure found between these domains can be seen as arising from the first two steps in a hierarchy proposed by the homotopy type theory program. In the third section, I return to my groupoid case study to see what further elaborations of the concept have been made. From groupoids I then turn to the mathematical concept of space, and show again how it has stayed true to a basic idea of everyday map-making, representing a space through overlapping charts. I briefly also consider the concept of duality.

VARIETIES OF CONCEPT

The term 'concept' itself designates a concept of course, one for which we should not expect a neat, formal definition. It belongs in English to a cluster of terms, including 'idea' and 'notion'. Whether these are used as synonyms or as slight variants is not easy to settle. In my book (Corfield, 2003) I tended to rely overly on the term 'idea', as acutely observed by the mathematician Michael Harris.

> Corfield uses the same word to designate what I am calling 'ideas' ('the ideas in Hopf's 1942 paper', p. 200) as well as 'Ideas' ('the idea of groups', p. 212) and something halfway between the two (the 'idea' of decomposing representations into their irreducible components for a variety of purposes, p. 206). Elsewhere the word crops up in connection with what mathematicians often call 'philosophy,' as in the 'Langlands philosophy' ('Kronecker's ideas' about divisibility, p. 202), and in many completely unrelated conections as well. Corfield proposes to resolve what he sees as an anomaly in Lakatos' methodology of scientific research programmes as applied to mathematics by a shift of perspective from seeing a mathematical theory as a collection of statements making truth claims, to seeing it as the clarification and elaboration of certain central ideas (p. 181).
>
> He sees 'a kind of creative vagueness to the central idea' in each of the four examples he offers to represent this shift of perspective; but on my count the ideas he chooses include two 'philosophies,' one 'Idea,' and one which is neither of these. (Harris, 2008, p. 16)

It is not that I believe I would have helped my case greatly by mixing things up with the use of, say, 'concept'. It seems to me that little is gained by carving up this semantic space with different terms, for example, by providing distinct, precise definitions for 'concept', 'idea' and 'notion'. But, as Harris suggests, there does seem to be a difference between, on the one hand, a core constituent of the visionary outlook of a large scale research programme, like that of Langlands, and, on the other, some stably defined kind of entity, such as a group. There is also a difference between the 'concept of (a) group' and the 'concept of symmetry', even if groups are the tool of choice to understand symmetry.

What we most need is as accurate as possible a description of mathematics as a tradition of intellectual enquiry. Any distinctions we draw should follow those we find in such a description. As Dudley Shapere (1987) argued, philosophers should not look to produce descriptions of science by refining 'meta-scientific' terms such as observation, confirmation and explanation independently from science itself, while the scientists go about their business, finding out about electrons, genes and tectonic plates. What it is to be an observation, a confirmation or an explanation changes as science learns more about the world and learns how to learn more effectively about the world. Similarly, it is not for philosophers of mathematics to impose some preordained notion of mathematical concept onto an account of mathematics. Indeed, we should expect such a 'meta-mathematical' term to require modification as mathematics proceeds.

One way we might expect to see these changes is through our foundational languages, especially when such a change occurs because of the demands of the practice. Certainly one of the most important motivations for Frege's work on his *Begriffsschrift* was to allow for the isolation and formation of fruitful concepts (Tappenden, 1995). The components of first-order logic – variables, function symbols, relation symbols, connectives and quantifiers – thus provide a range of tools. With the rise of set theory, to these were added the resources of the axioms of set formation. However, in recent decades doubts have grown as to how helpful these tools have proved to be in making sense of the details of mainstream mathematics. It has been a common observation that while everything is in principle expressible in set theory, to carry out such expression would provide no illumination.

With the emergence of a new foundation, homotopy type theory, there should be many opportunities to explore its potential for concept formation. Type theories have a large array of ingredients, from type formation rules, such as those for *identity* types and *inductive* types, to introduction and elimination rules for terms. We should be able to see then that some concepts emerge from the intrinsic components of the system, while others are merely expressible within it. Type theories have very close ties with category theory (Harper, 2013), which has a greater claim to represent faithfully, if not all of mainstream mathematics, then certainly large parts, such as algebraic geometry and algebraic topology. Homotopy type theory takes this association further to higher category theory (Lurie, 2009a). In addition, some type theories allow for the direct expression of 'modalities,' to be mentioned in the fifth section of this chapter, which allow forms of variation, cohesion and smoothness to be expressed. There are also linear flavours of type theory, associated with linear algebra and quantum mechanics (Schreiber, 2014). There will be interesting opportunities to find out which concepts are deeply ingrained in such type theories.

Since it is all too easy for philosophers to be seduced into believing that their preferred formalism is the source of all important answers, it will help to keep us honest to pay close attention to the best thinking about concepts in the philosophical literature, especially that tuned to scientific concepts. Mark Wilson in *Wandering Significance* (2006) presents a strong case that concepts are rarely given by neat necessary and sufficient conditions, and that they rarely map straightforwardly onto the world. Instead, concepts are patched together from varied parts. Different patches provide different 'directivities' as to application. We tend to cover up these differences, he claims, behind 'classical facades.' When we come to see through such facades, Wilson argues, we ought not retreat to what he calls a 'hazy holism.'

Many of Wilson's examples come from applied mathematics and engineering, where we hear of the wandering significance of hardness, force, weight and gearwheel. The imagery used for concept extension is mathematical: prolongation, lifting, pulling back. Mathematical concepts are included too, as we hear in his description of Dedekind's number theoretic analogy, the largest case study of chapter 4 of Corfield (2003). In future work I will explore whether we might take Wilson's mathematical imagery of concept extension more literally when applied in mathematics itself. After all, the terms he deploys to convey this imagery are generally spatial and may perhaps be better addressed by a foundational system tuned to spatial notions, such as homotopy type theory. But now we need to delve into a case study to start us off with examples of the kinds of material we need for the theses of this chapter. Let's then consider a simple analogy between logic and arithmetic.

ADDITION, MULTIPLICATION AND EXPONENTIATION

When I teach an introductory course on logic to philosophy students, I like to point out to them certain similarities between the truth tables for the logical connectives and some simple arithmetic operations, similarities which have been known since the nineteenth century. If we assign the values 1 to True and 0 to False, then forming the conjunction ('and') of two propositions, the resulting truth value is formed very much as a product of numbers chosen from {0, 1} is formed:

- Unless both values are 1, the product will be 0.
- Unless both truth values are True, the truth value of the conjunction will be False.

It is natural for students then to wonder if the disjunction ('or') of two propositions corresponds to addition. Here things do not appear to work out precisely. It is fine in the three cases where at least one of the propositions is False, for example, 'False or True' is True, just as $0 + 1 = 1$. But in the case of 'True or True', we seem to be dealing with an addition capped at 1, something such as would result in a third glass when the contents of two other full or empty glasses are emptied into it.

Setting to one side for the moment this deviation, the next connective usually encountered is 'implies'. A very good way of conveying to students the meaning of an implication between two propositions, such as $A \to B$, is as an 'inference ticket', a process taking a guarantee for the truth of A into such a guarantee for B. Implications seem to act very much like functions. Denoting the conjunction of A and B by $A \wedge B$, we may further observe the analogy:

- $(A \wedge B) \to C$ is True if and only if $A \to (B \to C)$ is True.
- $c^{(a \times b)} = (c^b)^a$

We may reformulate the former as expressing the equivalence between deducing C from the premises A and B, and deducing B implies C from the premise A.

Now, what accounts for this pleasant analogy? The idea of a proof of $A \to B$ as a function and its analogy to the arithmetic construction b^a should bring to mind the fact that the latter may also be interpreted as related to functions. Indeed, b^a counts the number of distinct functions from a set of size a to a set of size b, so that, for example, there are nine distinct functions from a set of size 2 to one of size 3.

Now we can see a closer analogy between proposition and sets, first through a parallel between conjunction and product:

- Guarantees of the proposition $A \wedge B$ can be formed from pairing together a guarantee for A and one for B.

- Elements of the product of two sets $A \times B$ can be formed from pairing together an element of A and one of B.

And second, there's a parallel between implication and function space, or exponential object:

- Guarantees of the proposition $A \rightarrow B$ are ways of sending guarantees for A to guarantees for B.
- Elements of the function space of two sets B^A (or $Fun(A, B)$) are functions mapping elements of A to elements of B.

This approach to logic is very much in line with the tradition of Per Martin-Löf (1984), and in philosophy with the ideas of Michael Dummett (1991). What homotopy type theory[1] does is take observations such as these not merely as an analogy but as the first steps of a series of levels. Propositions and sets are both considered to be simple kinds of *types*. In Martin-Löf style type theories, given a type and two elements of that type, $a, b: A$, we can form the type of identities between a and b, often denoted $Id_A(a, b)$. The innovation of homotopy type theory is that this identity type is not required to be a proposition – there may be a richer type of such identities. But we can see now how the levels bottom out. If you give me two expressions for elements of a given *set*, the answer to the question of whether or not they are the same is 'yes' or 'no'. We say that the identity type is a proposition. On the other hand, in the case of a type which is a proposition, were we to have constructed two elements, in this case proofs, and so know the proposition to be true, these elements could only be equal. Here we are not distinguishing between different proofs of a proposition, which has led some to call them 'mere propositions.'

So taking identity types of the terms of a type decreases the type's level, and this may be iterated until we reach a point where the type resembles a singleton point. For historical reasons connected to the branch of mathematics known as homotopy theory, this level is designated −2. Mere propositions are thus of level −1, and sets of level 0. But we can find higher-level types, ones for which there may be a *set* of identities between two terms, or a type of identities of yet higher level. Take the type of finite sets. For any two elements, that is, finite sets, an element of their identity type is any map which forms a one-to-one correspondence between the sets. In the case of large equally sized sets, there will be many of these correspondences. The

[1] The term 'Univalent Foundations' is sometimes used synonymously, possibly with a slight difference of emphasis. For a textbook treatment, see Univalent Foundations Program (2014).

identity type is a set. But then there is nothing further to say about any two such identities, other than that they are the same or different. This type of finite sets therefore has level +1, since identity types are sets, of level 0. And so it goes on. One may have types of infinitely high level.

An important part of Martin-Löf type theory is the notion of a dependent type, denoted $x : A \vdash B(x)$ *Type*. Here the type $B(x)$ *depends* on an element of A, as in *Days*(m) for m: *Month*, or *Players*(t) for t: *Team*. Two useful constructions we can apply to these types are *dependent sum* and *dependent product*.

- Dependent sum: $\Sigma_{x:A} B(x)$ is the collection of pairs (a, b) with a: A and b: $B(a)$. When A is a set and $B(x)$ is a constant set B, this construction amounts to the product of the sets. Likewise if A is a proposition and $B(x)$ is a constant proposition, B, dependent sum is the conjunction of A and B. In general we can think of this dependent sum as sitting 'fibred' above the base type A, as one might imagine the collection of league players lined up in fibres above their team name.
- Dependent product: $\Pi_{x:A} B(x)$, is the collection of functions, f, such that $f(a)$: $B(a)$. Such a function is picking out one element from each fibre. When A is a set and $B(x)$ is a constant set B, this construction amounts to B^A, the set of functions from A to B. Likewise if A is a proposition and $B(x)$ is a constant proposition, B, dependent product is the implication $A \rightarrow B$. In terms of the picture of players in fibres over their teams, an element of the dependent product is a choice of a player from each team, such as *Captain*(t).

Here then we can see how close this foundational language is to mainstream mathematics and physics. Dependent sum and dependent product correspond, respectively, to the total space and to the space of sections of fibre bundles, which appear in gauge field theory often in the guise of principle bundles. A fibre bundle is a form of product, but a *twisted* one where as we pass around the base space, a fibre may be identified under a nontrivial equivalence. An easy example is the Moebius strip, where an interval is given a 180 degree twist as it is transported around a circle. Of course, for physics one needs some geometric structure on the base space and total space. We shall see more about this in the fifth section of this chapter.

We can now see that these constructions allow us to formulate the quantifiers from first-order logic, at least as defined over types. So consider the case where A is a set, and $B(a)$ is a proposition for each a in A. Perhaps A is the set of animals, and $B(a)$ states that a particular animal, a, breathes. Then an element of the dependent sum is an element a of A and a proof of $B(a)$, so something witnessing a breathing animal. Meanwhile an element of the dependent product is a mapping from each $a: A$ to a proof of $B(a)$. There will only be such a mapping if $B(a)$ is true for each a. If this were the case, we would have a proof of the universal statement 'for all x in A, $B(x)$', in our example, 'All animals breathe.'

Returning to the dependent sum, this is almost expressing the existential quantifier 'there exists x in A such that $B(x)$', except that it is gathering all such a for which $B(a)$ holds, or, in our case, gathering all breathing animals. As we have seen before in the capped addition of the Boolean truth values in a disjunction, to treat this dependent sum as a proposition, there needs to be a 'truncation' from set to proposition, so that we ask merely whether this set is inhabited, in our case, 'Does there exist a breathing animal?' This extra step should be expected as existential quantification resembles forming a long disjunction. That we do not need to adapt for universal quantification tallies with the straightforward form of the product of Boolean values.

What emerges from these lines of thought is that the lower levels of homotopy type theory have contained within them: propositional logic, (typed) predicate logic and a structural type theory. Coming from a tradition of *constructive* type theory, one needs to add classical axioms if these are required, such as various forms of excluded middle.

Before ending this section, I shall introduce a further way to think about dependent sum and dependent product, this time as related via *adjoints*. Adjunctions are category theory's way of dealing with something as close to an inverse as possible when such an inverse does not necessarily exist. Then left and right adjoints are approximations from either side.

Returning to our type theory, given a plain type, A, we can turn any type C into one dependent on A by formulating $x : A \vdash (A \times C)(x) := C$. If A and C are sets, think of lining up a copy of C over every element of A, the product of the two sets projecting down to the first of them, $A \times C \to A$. Now, we can think of approximating an inverse to this process, which would need to send A-dependent types to plain types. Such approximations, or adjoints, do indeed exist. Left adjoint to this mapping is dependent sum, and right adjoint is dependent product. The fact that these are adjoints may be rendered as follows, for B a type depending on A:

- $Hom_C\left(\sum_{x:A} B, C\right) \cong Hom_{C/A}\left(B, A \times C\right)$
- $Hom_C\left(C, \prod_{x:A} B\right) \cong Hom_{C/A}\left(A \times C, B\right)$

I shall not further explain this formulation except to note that we can see vertical unity at play here if we return once again to the logic-arithmetic analogy from earlier in this section. Take *A*, *B* and *C* as finite sets, with *B* fibred over *A*, which just amounts to a map from *B* to *A*, the elements of *B* fibred above their images in *A*. Then taking the cardinalities of the two sides of each adjunction above yields further recognisable arithmetic truths:

- $C \sum_i b_i = \prod_i C^{b_i}$
- $\left(\prod_i b_i\right)^C = \prod_i (b_i)^C$

So a pupil being taught that, say, $3^5 \times 7^5 = 21^5$ is being exposed to the shadow of an instance of an important adjunction, which in turn, as with all of the preceding discussion of dependent sums and products, works for types up and down the hierarchy of *n*-types. Here *vertical unity* takes on a literal meaning. It is these *n*-types in the guise of higher groupoids that we turn to next.

GROUPOIDS, IDENTITY AND SYMMETRY

As it has turned out, my choice of concept in Corfield (2001) was a good one. Groupoids have flourished in the intervening years, and we can now update their history a little to say more about the hierarchical levels of homotopy type theory. A central case for groupoids that I discussed was their ability to capture 'irregular' quotients. This can be illustrated by a simple example which I first learned from John Baez (see Baez & Dolan, 2001). Consider a row of six objects spaced evenly along a line (Figure 7.1). Inverting the line about the midpoint identifies the objects, yielding three pairs.

There is nothing surprising here. We can take ourselves to be demonstrating the simple mathematical fact that 6/2 = 3. Whether we take this to be three bound pairs or three singletons matters little. Now let us do the same thing with five objects (Figure 7.2). In this case the central object is unlike the others in that it will be paired with itself.

Classes formed by the identification number three: two pairs and that odd singleton. Here it seems we could be construed as making the evidently false claim that 5/2 = 3. To repair this, it would appear that we should have that special central object count only for one-half. Groupoids give us a way to do precisely this.

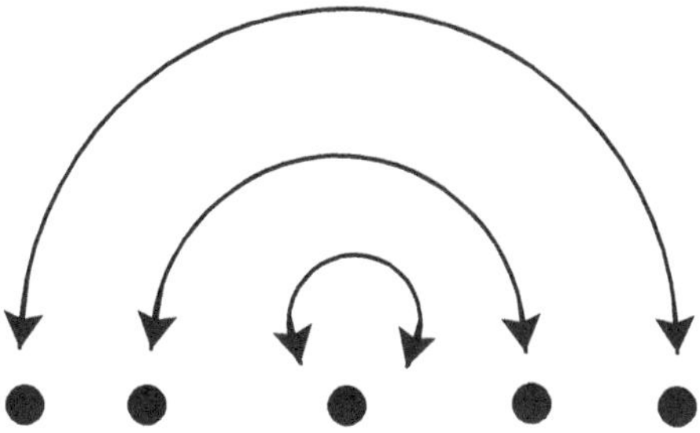

FIGURE 7.1 Halving 6.

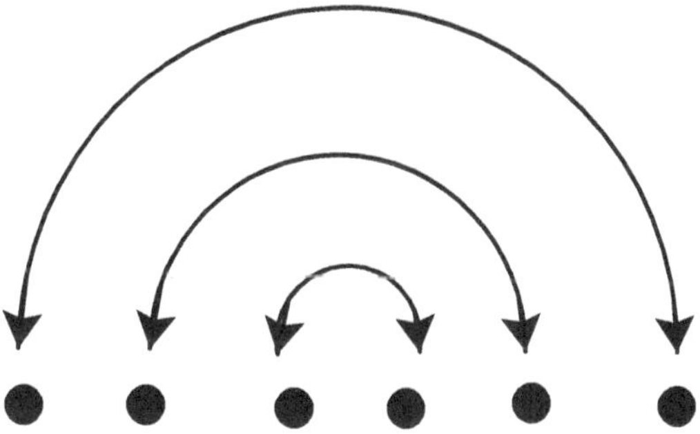

FIGURE 7.2 Halving 5.

For any group, in the mathematical sense, acting on a set, we speak of the 'action' of the group, and represent it by the 'action groupoid'. Think of the elements of the set as points or objects, and arrows going from one object to another are labelled by group elements which send the first object to the second. The cardinality of a groupoid is found by summing for each 'component', or connected cluster of objects, the reciprocal of the size of its 'isotopy group', essentially the number of arrows looping from an object to itself. In the case of five objects above, for the object identified to itself in the middle of the row, there are two such arrows. In this case, we have to add $1+1+1/2=5/2$, as desired. Now it is a general result that whatever the group, and whatever its action on a finite set, we will have an arithmetic identity in that the cardinality of the action groupoid is the size of the finite set divided by the order of the group.

For the purposes of physics, again, we need more geometric structure, but the basic idea of an action by a group of transformations specifying invariance under symmetries is at work. From 'orbifolds', which might result from a Lie group acting unevenly on a manifold, to the objects of field values for gauge field theory, which need to keep track of gauge equivalences, we find we are treating geometric forms of groupoid. If in the latter case instead we take the simple quotient, the plain set of equivalences classes, which amounts to taking gauge symmetries to be redundant, the physics goes wrong in some sense, in that we cannot retain a *local* quantum field theory. Furthermore, in the case of gauge equivalence, we do not just have one level of arrows between points or objects, but arrows between arrows and so on, representing equivalences between gauge equivalences. And this process continues indefinitely to 'higher' groupoids, needed to capture the higher symmetries of string theory (Schreiber, 2013). In conceptual terms, the very idea of sameness or identity is being given a much richer interpretation here (Corfield, 2003, chap. 10). This is embodied in the form of the identity types we saw in the previous section.

The action groupoid is in fact a manifestation of something we have already seen. Indeed, in homotopy type theory, the way to express that a set, A, is acted on by a group, G, is to write it as a dependent type $* : BG \vdash A(*)$: *Type*. The type BG is a version of the group G, but where we are taking it to be a single object with looping arrows labelled by each group element. Applying the dependent sum construction from the last section, we find that $\Sigma_{*:BG} A(*)$ is a type with structure that of the action groupoid, which should indicate to us that the concept is a fundamental one.

Similarly we can form the dependent product, which in this case is composed of the 'fixed points' of the action, those elements of the set left unchanged by all elements of the group. Carrying on in this vein, we soon meet some central constructions from the area known as 'representation theory', a subject taught at the earliest at a senior undergraduate level. So it turns out that important concepts in representation theory are just constructions in homotopy type theory carried out in a context of symmetries, constructions which in more basic settings are familiar logical operations. For example, the type theoretic construction governing the formation of *induced representations*, typically first met as a way of expanding an action of a subgroup to an action of the full group, is very basic when carried out on sets and predicates. For instance, consider the case of the set of dogs sent by the function 'owner' to the set of people, each dog being assigned its (unique) owner. The equivalent of forming an induced representation in this setting is to map any predicate of dogs to a predicate of people, say, 'poodle' is sent to 'owner of some poodle'. Again we find this new foundational language showing us vertical unity between the topic of an advanced undergraduate or graduate course and something the general populace might understand.

With such little conceptual guidance provided by set theory, it is unsurprising that we turn to the naming of concepts for illumination, even if this may be misleading. It seems that there is a thin line between modifying the characterisation of a concept and devising a new neighbouring concept. Sometimes we just add an epithet to an existing term, as in *imaginary* number or *nondifferentiable* function. Sometimes we modify one, such as *functor* from *function*. However, naming choices are not strictly dictated by how things are best understood. Indeed, there is no reason to take preservation of a name as a sign of great stability. Concerning the natural sciences, Dudley Shapere (1989) once made the excellent point that we can be taken in by the vagaries of naming decisions to see a larger or smaller difference than we ought. He gave the case of 'atom'. Because the name has been kept, we strive to find much greater continuity than for the case of a concept which has changed similarly but been renamed. In fact, in the early nineteenth century 'atom' and 'molecule' were used interchangeably. Had the latter been the only term, a story of the change from 'molecule' then to 'atom' now should involve no more of a change than we tell now from 'atom' then to 'atom' today. It has been argued (Chang, 2009) that theories of phlogiston had comparable levels of empirical adequacy to early theories of oxygen, and that judgements as to the justified triumph of the latter are largely due to its having won out historically in the early nineteenth century. We pay the price today by having retained the poorly named oxygen, the supposed 'principle of acidity'.

Over time there is an inclination to absorb the earlier term into a generalisation by dropping the epithet. Then we may need a new term to designate what was special about the original case. Think of 'natural' number or 'differentiable' function. On the other hand, sometimes we resist such terminological change as with 'multivalued function', which is still not an example of a function. This is sometimes referred to as a 'red herring'. In the case of groups, in the early days of their use, Poincaré meant by 'group' what we call *abelian* group. By the time Brandt coined 'groupoid' in the 1920s, the term 'group had the settled usage we see today. He might quite possibly have denoted one as a '*something* group', rather than use the '-oid' suffix. Having opted for 'groupoid', it was always going to be hard to drop the suffix. Now that we have higher groupoids and their specialisation to higher groups, there is perhaps a chance to simplify the terminology. But rather than naming policies, the important point of view provided by homotopy type theory reveals the centrality of this cluster of concepts to a conception of mathematics based on identity and structural invariance.

SPACES AND DUALITY

In this final section I would like to discuss briefly topics of two further studies of mine (2015, forthcoming) which concern the large thematic concepts of space and duality. A common criticism of set theory as a foundational language has been its unnatural treatment of certain concepts, none more so than spatial ones. Where the set theoretic approach can only look to build up a set that behaves like a space from the dust of its points, homotopy type theory can at least build into its basic entities something path-like in the collection of identities between terms. But we must be careful here as,

> although it is common in homotopy type theory to use terminology borrowed from topology such as 'path' and 'circle', these words have *a priori* nothing to do with their topological versions *which can also be defined inside of type theory.* (Shulman, 2015, p. 3)

Indeed, homotopy type theory is a synthetic theory of structure, not of spaces (Shulman, forthcoming). This point can be seen clearly from the two versions of circle available in the theory. The native one uses 'higher inductive types' to define a type, *S*, with a designated element known as its base point, *base*, along with a designated identity element in $Id_S(base, base)$. It is natural to represent this as a point with a loop which may be iterated indefinitely, just as one might wind a string around a circular reel. This type can then be used to do a synthetic form of algebraic topology, such as finding the type of maps from *S* to itself as an intrinsic form of the fundamental group of the circle, which is then calculated to be isomorphic to the integers (Licata & Shulman, 2013).

The second approach is to define a circle as a locus of points in the plane. This is a much more elaborate, and less intrinsic, construction in the type theory, which first requires the construction of the reals, before forming the plane and then the subtype of this which is the circle, in much the way one would ordinarily produce a circle. In the bare type theory there is no connection between these two 'circles'.

Intrinsic ways to define continuity or cohesiveness have therefore been sought, and a solution recently found in the form of the addition of some 'modalities' to homotopy type theory (Schreiber, 2013). These are expressed using a string of adjunctions in a way we need only hint at here. The treatment comes from an understanding of a classic case where topological spaces and sets are conceived as related by a series of four adjoint functors. The pair from spaces to sets send a space (a) to the set of its connected components, as though any points connected by a path had been clumped together to a point, and (b) to the set of its points, as though any connectivity between points had been dissolved. The two functors in reverse send a set to the spaces with that underlying set and either the discrete or codiscrete topology, so either one applies no glue or a maximal glue when forming the spaces. These four functors can be composed in three different ways to yield adjoint functors from spaces to spaces. Abstracting from this situation, a pattern of such a triple of adjunctions gives a general way to express how something spatial is made of parts which hold together cohesively. Now we can relate the two circles above. One can also add in a further triple of adjoints to allow the synthetic expression of many major constructions from differential topology and geometry, to capture smoothness as well as continuity.

As I explain in Corfield (forthcoming), however intricate the 'spaces' of current mathematics, we never leave behind the essential idea of constructing them by gluing together specified model spaces. This is recognised by the term 'atlas' used for a collection of 'charts' which cover a manifold, but is an idea present in the recent 'generalised schemes' of Jacob Lurie (2009b) and other spaces of current geometry. This patchwork idea of spaces is perhaps a feature of our everyday understanding of our surrounding space. It seems that when we work out a way to walk across a city, we have something like local maps in our mind, which we can look to adjoin to provide us with a feasible route. With the addition of the modalities, homotopy type theory can express this gluing process straightforwardly, as it can the geometric constructions needed to formulate gauge field theories.

Finally, we turn to duality, which in some ways presents an interesting problem for Thesis 1 in that it runs in some sense orthogonal to the n-type hierarchy, and so may fit less well with what we have discussed so far. That it is a pervasive thematic mathematical concept is clear, and that it is superbly well captured by category theory is readily established (Corfield 2015). Moreover, concerning the vertical unity of Thesis 2, duality is certainly rooted in simple, everyday concepts. Take a cube and join the midpoints of its adjacent faces with edges to form an octahedron. Then do the same with an octahedron to form a cube. The eight vertices, twelve edges and six faces of the cube are dual to the eight faces, twelves edges and six vertices of the octohedron. We can likewise pair the dodecahedron with the icosahedron and the tetrahedron with itself to complete the platonic solids. This is an example of what is sometimes called *formal* or *abstract duality*, where there is an exchange of kinds of entity and relations between them.

Important pieces of mathematics then arise from relating this formal duality with *concrete* duality, which arises from mappings which assign a value to pairs of entities. Consider a simple truth theoretic pairing, where for an actor and a film we assign a truth value depending on whether or not they appear in the film. For a typical range of actors and films it will be the case that the list of films in which an actor appears determines that actor and the list of actors in a film determines that film. In mathematics we often find *perfect* pairings between two collections where it is possible to reconstruct an element of the first collection from the values resulting by pairing it with elements of the second collection, and vice versa. The inner product of a vector space is an example of this situation where a vector is determined by the values given by its inner product with a complete set of linearly independent vectors. The duality between platonic solids can be reconsidered in this setting, relating formal and concrete dualities.

While vertical unity is apparent within the topic of duality, stretching from these simple examples to the very forefront of mathematical research, the issue of how it relates to varieties of type theory is not so clear yet. Naturally, one can reconstruct specific examples of duality in these languages, just as one can reconstruct any mathematics within set theory, but one might expect general aspects of such a fundamental concept to be expressible synthetically. However, the research front is moving fast here. It appears that there is a *linear* version of homotopy type theory, which captures some of the duality structures found in the mathematics of quantum physics (Schreiber, 2014), and elsewhere proposals for an 'adjoint logic'. It will be interesting to see how this work plays itself out.

CONCLUSION

In this chapter I set out to make the case for two theses:

1. This new foundational language, homotopy type theory, provides an important perspective from which to understand varieties of mathematical concept.
2. Mathematics displays *vertical unity*, so that concepts met in elementary and in research-level mathematics are related.

As I remarked, these theses are related in that the new language is often able to reveal the commonality between elementary and advanced concepts as arising from a single form of construction within the type theory.

At the same time, there is no claim on my part that every mathematical concept finds its simplest expression within homotopy type theory. However, we should see here the possibility of assessing constructions as to whether they are intrinsic to all type theories, intrinsic to specific type theories (such as the homotopy, modal homotopy and linear varieties) or as requiring complex definitions within a specific type theory. The product construction, manifesting as set product and conjunction of propositions, is a good example of a very basic form of concept.

Eventually we may find some form of complexity measure for a concept or result which adds the complexity of a formal framework to that of the construction of the concept or result within that framework. We should expect that what is complicated according to one type theory may be approached more intrinsically within another. Good examples of this are Stokes' theorem and Noether's theorem which appear highly complex in set theory, but which become near-logical results when expressed in modal homotopy type theory (Schreiber, forthcoming). In that homotopy type theory and other type theories fare well at displaying the reasons for vertical unity, I believe we can take this as a good indication of their power.

REFERENCES

Baez J. & Dolan J. (2001). From finite sets to Feynman diagrams. In B. Engquist and W. Schmid (Eds.), *Mathematics Unlimited – 2001 and Beyond, vol. 1*, (pp. 29–50) Berlin: Springer. arXiv preprint: http://arxiv.org/abs/math/0004133.

Borovik, A. (2005). What is it that makes a mathematician?, www.maths.manchester.ac.uk/avb/pdf/WhatIsIt.pdf

Chang H. (2009). We have never been whiggish (about phlogiston), *Centaurus*, 51(4), 239–264.

Corfield, D. (1997). Assaying Lakatos's philosophy of mathematics. *Studies in History and Philosophy of Science Part A*, 28(1), 99–121.

(1998). Beyond the methodology of mathematics research programmes. *Philosophia Mathematica*, 6(3), 272–301.

(2001). The importance of mathematical onceptualisation. *Studies in History and Philosophy of Science Part A*, 32(3), 507–533.

(2003). *Towards a philosophy ofreal mathematics.* Cambridge: Cambridge University Press.

(2015). Duality as a category-theoretic concept. *Studies in the History and Philosophy of Modern Physics*, doi:10.1016/j.shpsb.2015.07.004

(forthcoming). Reviving the philosophy of geometry. In E. Landry (Ed.), *Categories for the working philosopher.* Oxford: Oxford University Press.

Dummett, M. (1991). *The logical basis of metaphysics*. London: Duckworth.

Harper, R. (2013). Computational trinitarianism. www.doc.ic.ac.uk/gds/PLMW/harper-plmw13-talk.pdf

Harris, M. (2008). Why mathematics you may ask? In Gowers, T. et al. (Eds.), *Princeton companion to mathematics* (pp. 966–977). Princeton, NJ: Princeton University Press.

Licata, D. & Shulman, M. (2013). Calculating the fundamental group of the circle in homotopy type theory. In *LICS 2013: Proceedings of the Twenty-Eighth Annual ACM/IEEE Symposium on Logic in Computer Science*, pp. 223–232.

Lurie, J. (2009a). *Higher Topos theory*. Princeton, NJ: Princeton University Press.

(2009b). Derived algebraic geometry V: Structured spaces, arXiv preprint: http://arxiv.org/abs/0905.0459

Martin-Löf, P. (1984). *Intuitionistic type theory*. Napoli: Bibliopolis.

Schreiber, U. (2013). *Differential cohomology in a cohesive-topos*, arXiv preprint: http://arxiv.org/abs/1310.7930.

(2014). Quantization via linear homotopy types, arXiv preprint: http://arxiv.org/abs/1402.7041.

(forthcoming). Higher prequantum geometry. In G. Catren and M. Anel (Eds.), *New spaces in mathematics and physics*.

Shapere, D. (1987). Method in the philosophy of science and epistemology. In Nancy J. Nersessian (Ed.), *The process of science: Contemporary philosophical approaches to understanding scientific practice* (pp. 1–39). Dordrecht: Martinus Nijhoff.

(1989). Evolution and continuity in scientific change. *Philosophy of Science*, 56(3), 419–437.

Shulman, M. (2015). Brouwer's fixed-point theorem in real-cohesive homotopy type theory, arXiv preprint: http://arxiv.org/abs/1509.07584.

(forthcoming). Homotopy type theory: A synthetic approach to higher equalities. In E. Landry (Ed.), *Categories for the working philosopher*. Oxford University Press. arXiv preprint: http://arxiv.org/abs/1601.05035.

Tappenden, J. (1995). Extending Knowledge and 'Fruitful Concepts': Fregean Themes in the Foundations of Mathematics. *Nous*, 29(4), 427–467.

Univalent Foundations Program (2014). *Homotopy type theory*.

Wilson, M. (2006). *Wandering significance*. Oxford: Oxford University Press.

8

The Perfectoid Concept: Test Case for an Absent Theory

MICHAEL HARRIS

PERFECTOID PROLOGUE

It is not often that contemporary mathematics provides such a clear-cut example of concept formation as the one I am about to present: Peter Scholze's introduction of the new notion of *perfectoid space*. The twenty-three-year-old Scholze first unveiled the concept in the spring of 2011 in a conference talk at the Institute for Advanced Study in Princeton. I know because I was there. This was soon followed by an extended visit to the Institut des Hautes Études Scientifiques (IHES) at Bûres-sur-Yvette, outside Paris – I was there too. Scholze's six-lecture series culminated with a spectacular application of the new method, already announced in Princeton, to an outstanding problem left over from the days when the IHES was the destination of pilgrims come to hear Alexander Grothendieck, and later Pierre Deligne, report on the creation of the new geometries of their day. Scholze's exceptionally clear lecture notes were read in mathematics departments around the world within days of his lecture – not passed hand-to-hand as in Grothendieck's day – and the videos of his talks were immediately made available on the IHES website. Meanwhile, more "killer apps"[1] followed in rapid succession in a series of papers written by Scholze, sometimes in collaboration with other mathematicians under thirty (or just slightly older), often alone. By the time he reached the age of twenty-four, high-level conference invitations to talk about the uses of perfectoid spaces (I was at a number of those too) had enshrined Scholze as one of the youngest elder statesmen ever of *arithmetic geometry*, the branch of mathematics

[1] This expression was used on my blog by Mike Shulman in talking about Homotopy Type Theory: "I personally believe that the 'killer apps' of HoTT are likely to lie not in classical homotopy theory but in higher topos theory, where the distance between HoTT and the classical tools is greater and the benefits are more pronounced."

where number theory meets algebraic geometry. Two years later, a week-long meeting in 2014 on Perfectoid Spaces and their Applications at the Mathematical Sciences Research Institute in Berkeley broke all attendance records for "Hot Topics" conferences.

Four years after its birth, *perfectoid geometry*, the theory of perfectoid spaces, is a textbook example of a *progressive research program* in the Lakatos sense. It is seen, retrospectively, as *the right* theory towards which several strands of arithmetic geometry were independently striving. It has launched a thousand graduate student seminars (if I were a historian, I would tell you exactly how many); the students' advisors struggle to keep up. It has a characteristic terminology, notation, and style of argument; a growing cohort of (overwhelmingly) young experts, with Scholze and his direct collaborators at the centre; a domain of applications whose scope continues to expand to encompass new branches of mathematics; and an implicit mandate to unify and simplify the fields in its immediate vicinity. Last but certainly not least, there is the generous, smiling figure of Peter Scholze himself, in the numerous online recordings of his lectures or in person, patiently answering every question until his questioner is satisfied, still just twenty-seven years old, an inexhaustible source of revolutionary new ideas.

For the historian of concepts (which I am not), there is only one problem with this picture: the concept of a perfectoid space is one of the most difficult notions ever introduced in arithmetic geometry, which has a long tradition of difficult notions. The reader of this essay faces a second problem: not only am I not a historian of concepts, I am by no means an expert on the kinds of mathematics that have been put together to create the new concept of perfectoid spaces. So I will have to skim the surface, at the risk of distortion; but even the surface will be unfamiliar to most readers. I hope that the technical details will nevertheless provide the reader with the means of perceiving how the somewhat extreme case of perfectoid spaces exemplifies typical aspects of mathematical concept formation: the reference to a historical background (in this case, the long practice of studying properties of equations in number theory by relating them to properties of geometric objects), the identification of a specific range of open questions in this historical context and the expectation that they would be solved by the introduction of new techniques of a specific type, and the reception and general acceptance of the new concept as *the right* one because it met expectations in spite of its many novel features and because it lent itself immediately to solving outstanding open problems.

STARTING POINTS

Perfectoid spaces stand at the crossroads where topology, Galois theory, and the study of equations by means of congruences meet. It is a busy crossroads, already occupied by *étale cohomology* and *crystalline cohomology*, both created by Grothendieck and his school; by the new number systems (*Fontaine rings*) created by Jean-Marc Fontaine in his program to devise a *mysterious functor*, a canonical operation relating the two kinds of cohomology; and by the many technical innovations introduced in order to complete Fontaine's program. The most relevant of these for Scholze's work is the *almost mathematics* of Gerd Faltings, which in turn builds on the foundational work of John Tate, ninety years old and still active. Tate is also the creator of *rigid analytic spaces*, the first of several successful attempts to create a *p-adic geometry*. Perfectoid geometry does not transcend these previous attempts, but rather extracts those of their features that are relevant to the problems Scholze set out to solve.

I do not expect readers to possess any of the technical vocabulary introduced in the previous paragraph. Each of the technical terms, however, can be seen as the precise analogue of a familiar notion from higher-dimensional geometry in Euclidean space; for the purposes of this essay, much of the meaning of Scholze's innovations can be understood by reference to these analogies. I will start with the way differential calculus encodes topology, and specifically with *Green's theorem* in calculus in two variables (as a topic in second year calculus, no other starting point is nearly as elementary). Green's theorem relates the line integral of a differential

$$fdx + gdy$$

on a simple closed curve *C* to the double integral of an expression involving the partial derivatives of *f* and g on the region *D* bounded by the curve. This breaks down when *D* is no longer a region bounded by a simple closed curve, but rather a region with holes (see Figure 8.1). The breakdown is an expression of the complicated *topology* of *D* – the presence of holes – but Green's theorem has a generalisation – a version of *Stokes' theorem* that accounts for the holes. The generalisation of these theorems to a manifold *M* of arbitrary dimension relates the topology of *M* – and more precisely, the *cohomology* of *M*, which is the way of keeping track of generalised holes – to the possibility or impossibility of integrating differential forms on *M*. For general (compact) manifolds, the relation is the statement of *de Rham's theorem*. When *M* is the space of complex solutions (in projective

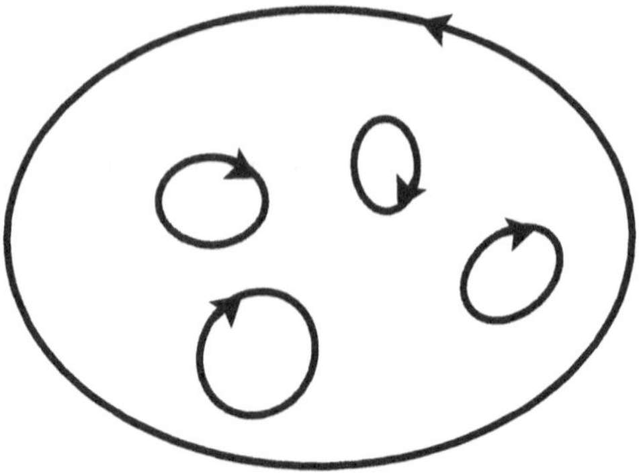

FIGURE 8.1 A region that contains holes and requires the generalisation of Green's theorem.

space) to a system of *polynomial equations* – a (smooth, projective, complex) algebraic variety – then *Hodge theory* gives a more precise relation, where the distinction between holomorphic differentials (like dz) and antiholomorphic differentials (like $d\bar{z}$) is central. For the purposes of this discussion, the important fact about Hodge's theory is that it uses differential calculus to compute the topology, or more precisely the cohomology, of a complex algebraic variety *M*. Another fruitful way of saying the same thing is to say that *M* has *two kinds* of cohomology groups: the topological one (that keeps track of holes), often called *Betti cohomology*; and the analytic one, based on the algebra of differential forms, called *de Rham cohomology*. Then the *comparison theorem* of Hodge theory – at this stage it is really de Rham's theorem – says that these are two different ways of computing the same thing.

Now forget the algebraic variety *M* but keep the polynomial equations used to define *M*. What happened to the holes? In the 1960s, Grothendieck and his collaborators constructed a purely algebraic invariant[2] of an algebraic variety – not necessarily smooth, not necessarily projective, and over any coefficient field *F* – that recovered the topological (Betti) cohomology groups, in the process transforming the notion of *topological space* practically beyond recognition. Something was gained and something was lost.

[2] André Weil had constructed such an invariant for varieties of dimension 1, speculated that such a theory should exist in general, and conjectured some of its desirable properties. Grothendieck proved the Weil conjectures, with the exception of the last one, which was solved in 1973 by Pierre Deligne. I was at the IHES, not yet a graduate student, the day the twenty-nine-year-old Deligne announced his solution. It was seen, correctly, as a very big deal.

What was gained was that the cohomology groups acquired an action of the Galois group of the coefficient field *F* – and when *F* is the field of rational numbers, or a finite field, this defines an invariant of exceptional richness. The *Tate Conjecture*, which has been proved in only a few situations (notably by Tate and Faltings), asserts that this Galois representation encodes a good deal of the information about solutions of the polynomial equations and higher-dimensional generalisations of solutions (whatever one might mean by that). What was lost is that, whereas Betti cohomology groups actually count holes (more precisely, they are finitely generated abelian groups), the algebraic invariants introduced by Grothendieck and his collaborators, the *étale cohomology* groups, only count holes modulo N for all integers N – with the important proviso that N is *not divisible by the characteristic of the coefficient field F.* It is more useful to package this information by prime powers and to look at the l-adic (étale) cohomology, which puts together the étale cohomology modulo l^n for all *n*, where l runs through the set of prime numbers that are invertible in the field *F*.[3] The underlying assumption is that the number of holes in an algebraic variety, or even a two-dimensional manifold, is a fundamental geometric property – an invariant.

Let us pause to consider that set of prime numbers l invertible in a fixed field F, since this is the most salient property of the field, and because the raison d'être of Scholze's work is to answer questions about equations over *p*-adic fields (where some standard methods of algebraic geometry do not apply) by "tilting" the equations so they are now over fields of characteristic *p*. In fact, the *characteristic* of *F* is decreed to be either the set of primes that are *not* invertible in *F*, if there are such, or else zero, if all primes are invertible in *F*. A field is of characteristic zero if and only if it contains the rational field **Q**; for example, the familiar fields of real or complex numbers are of characteristic zero. Otherwise, there is exactly one prime, traditionally denoted p, which is not invertible in *F*, and *F* is said to be of (positive) characteristic *p*. Since every non-zero element of *F* is invertible, this means that $p = 0$ in *F*, and F contains the *prime field* $\mathbf{F}_p = \mathbf{Z}/p\mathbf{Z}$ of *p* elements, the field whose arithmetic corresponds to the study of congruences modulo *p*. Scholze's initial motivation, very roughly speaking, was to study equations over certain kinds of fields of characteristic zero by reinterpreting them as equations over fields of prime characteristic *p*, where other methods are available.

[3] For a philosophically enlightening introduction to the topological approach to number theory, see §3 of McLarty (2010).

A field F of characteristic $p > 0$ has two closely related properties that it does not share with fields of characteristic zero.

(1) First, polynomials of positive degree can have derivative zero: indeed, if $P(X) = X^p$, then $P'(X) = pX^{p-1} = 0$ because $p = 0$.[4]
(2) On the other hand, the map *Frob* (for Frobenius) that sends $a \in F$ to its p-th power $Frob(a) = a^p$ is a *homomorphism* of rings:

$$Frob(ab) = Frob(a)Frob(b) \text{ and } Frob(a+b) = Frob(a) + Frob(b)$$

for any $a, b \in F$. The multiplicative property is obvious, the additive property follows from the binomial formula for $(a+b)^p$, because all the intermediate binomial coefficients are divisible by p. Because F is a field, only 0 has p-th power zero, and thus *Frob* is injective. The field F is called *perfect* if *Frob* is also surjective – in other words, if every element of F has pth roots (and therefore p^nth roots for all n). Scholze's term *perfectoid* is derived from this property of perfect fields.

I return now to the discussion of algebraic varieties over any field, which have differential forms, as in our earlier discussion. There is also an algebraic version of de Rham cohomology (whose main properties were outlined in a letter from Grothendieck (1966) to Atiyah). But to work properly with differentials of degree n, one needs to be able to divide by $n!$, and this is impossible if the characteristic of F is a prime dividing n; alternatively, property (1) above shows that one cannot integrate the differential $X^{p-1}dX$. Expanding the notion of space a bit more, Grothendieck outlined the properties of yet another algebraic cohomology theory (for smooth projective varieties), *crystalline cohomology*, whose construction was carried out in the thesis of his student Pierre Berthelot and his collaborators. It was around this time that Grothendieck abandoned the IHES and the mathematical community, but not before he had asked for an analogue of Hodge theory in the algebraic setting. The original Hodge theory was defined for varieties with complex coefficients; Grothendieck expected an analogue for varieties over *p-adic local fields*. For our purposes a field F is p-adic local if it is a topological field with an open subring $O = O_F$ such that F is obtained from O by allowing division by p, and such that O is a local ring (it has a unique maximal ideal m) and a discrete valuation ring (every ideal is a power of m such that $k = O/m$ is a finite field). When F is a p-adic local field, the algebraic de

[4] Didn't I just say that $p > 0$? So how can I say $p = 0$? The prime number p is and remains a positive number in the field of rational numbers, but for the purposes of arithmetic in a field of characteristic p it is treated as if it were equal to 0, and indeed this is the meaning of characteristic p. There is no logical inconsistency.

Rham cohomology, crystalline cohomology, and *p*-adic étale cohomology all have coefficients in the same kind of field, namely either *F* or a p-adic local field closely related to *F*. Grothendieck asked for a "mysterious functor" that played the role of the de Rham and Hodge theorems in relating the *p*-adic étale cohomology of an algebraic variety over *F* to its algebraic de Rham cohomology, and (when this makes sense) to the crystalline cohomology of its reduction over *k*. A few years earlier, Tate had introduced what are now called *Hodge-Tate structures* and, together with Jean-Pierre Serre, had formulated a different conjectural *p*-adic version of the Hodge theorem; Grothendieck's proposal would necessarily imply Tate's conjecture (not to be confused with "the Tate conjecture" mentioned earlier).

Fontaine introduced his new rings, including the ones called B_{dR} and B_{cris}, as a way of relating the Galois structure and the differential structure, and used them to define a candidate for Grothendieck's mysterious functor. There ensued an international contest, lasting nearly two decades, to prove that Fontaine's approach provided the mysterious functor. Along the way, the field saw the growth of the new research program of *p-adic cohomology*, characterized by the proliferation of new cohomology theories with p-adic coefficients, including *syntomic cohomology*, *de Rham-Witt cohomology*, and *rigid cohomology*, and centered around the problem of realizing Grothendieck's mysterious functor.[5]

The main goal of the program was to find better ways to understand the Galois representations constructed by Grothendieck, because they are among the central objects of algebraic number theory. Results of this research program were of fundamental importance in Andrew Wiles' proof of Fermat's Last Theorem and in subsequent work that derived from that of Wiles. Especially in the hands of Richard Taylor and his students and collaborators, this line of research has developed into one of the most active research programs in algebraic number theory, overlapping with the *Langlands program* on automorphic forms – and only incidentally in geometry. This is largely responsible for the growing familiarity of number theorists with Fontaine's *p*-adic constructions in the years preceding Scholze's perfectoid announcements.

A word about Fontaine's rings is in order, because of their importance for the theory of perfectoid spaces. These rings, which can be thought of as alternative *p*-adic analogues of the complex number field, are constructed by an elaborate series of steps,[6] of which the first invariably involves taking

[5] It was mysterious because Grothendieck did not know how to define it but assumed it must exist. Then Fontaine defined a functor, and over the years he and others showed that it had the expected properties.

[6] See, for example, http://en.wikipedia.org/wiki/Ring_of_p-adic_periods.

p^nth roots for all n in a field of characteristic zero in order to define a new and perfect field of characteristic p. In retrospect, this way of moving from characteristic 0 to characteristic p can be seen as a prototypically perfectoid thing to do. As recently as fifteen years ago, any talk at a number theory seminar involving one of Fontaine's rings would normally begin with a reminder of the ring's construction. I never found this very enlightening, because it was never explained why this particular sequence of otherwise unmotivated steps yielded an interesting object – in other words, the construction did not illuminate the *concept* of the Fontaine ring. These days, the rings tend to be introduced without explanation, which I take to mean that the seminar audience has become sufficiently familiar with the rings not to need to be reminded that the speaker knows how they are put together. In practice, this means that a typical (algebraic) number-theorist-in-training is expected to know a few standard properties of Fontaine's rings; one might (or might not) want to say in this connection that the number-theorist-in-training has been initiated into one or more of the *language games* in which Fontaine's rings feature prominently. More to our purpose, this familiarity means that the community of participants in number theory seminars was prepared for Scholze's theory of perfectoid spaces by repeated exposure to the theory and use of Fontaine's rings, largely in connection with outgrowths of the ideas of Wiles and Taylor.

To put an end to this lengthy introduction to the background to Scholze's work, Fontaine's hope to have realized the mysterious functor was confirmed, not once but several times, giving rise to what is now generally known as *p-adic Hodge theory*. The first complete proof was obtained by Faltings; a rather different method, extending ideas due to Fontaine and William Messing, was developed by Japanese specialists, culminating in a second (and initially more complete) complete proof by T. Tsuji.[7] Two relevant features of the Faltings approach deserve to be mentioned here. The first feature is the introduction of what Faltings called *almost mathematics*,[8] which is (very roughly) a form of commutative algebra in which certain

[7] There is a third proof, due to Wiesława Nizioł, using a rather different circle of ideas; and more recently there have been other proofs, notably one based on perfectoid geometry due to Scholze himself.

[8] The first complete treatment is contained in Gabber and Ramero (2003). Faltings' (1988) proof was published earlier. The "certain kinds of error" mentioned in the next sentence can be likened to measurement errors in physics. It is as if one were using two different scales to make multiple measurements of the geometry of an algebraic variety; the scales can give slightly different measurements in the intermediate steps, but in almost mathematics they are treated as identical. It turns out that the final result is insensitive to this kind of error.

kinds of error are systematically ignored, and in which it is shown rigorously that they do not matter to the final proof of Fontaine's comparison. The second feature is that Faltings made systematic use of constructions involving the taking of p^nth roots for all n. Faltings said at the time that his constructions were inspired by Tate's original work on p-adic Hodge theory.

SCHOLZE'S PERFECTOID CONCEPT

Scholze's perfectoid spaces are, in the first place, *spaces*. The notion of space in algebraic geometry has evolved through several stages since André Weil introduced abstract varieties by the gluing together of affine algebraic varieties in his *Foundations of Algebraic Geometry*. The current understanding is based on Grothendieck's framework, developed systematically by Grothendieck and Dieudonné in the *Éléments de géométrie algébrique* (EGA), following Serre's introduction of sheaf-theoretic methods. A *space* in this setting then consists of a topological space (the "underlying topological space," generally not Hausdorff) together with the (sheaf of) functions defined on its open sets. I pause to mention that Grothendieck's reconceptualisation of topology involves a generalisation of the notion of "open set" that is radically at odds with the primitive intuition of continuity that general topology was designed to formalise, but that nevertheless works magnificently.

In p-adic geometry, the topology is somehow adapted to the topology of the p-adic numbers, and the functions take p-adic values. Perfectoid spaces are spaces in the setting of p-adic geometry, and more precisely p-adic *analytic* geometry. Like the p-adic numbers themselves, p-adic geometry is a form of geometry designed to study solutions of congruences modulo powers of p. The p-adic numbers themselves form a topological space, but it is totally disconnected, which means that the space can be broken continuously into arbitrarily small pieces. This is not suitable for geometry; there are too many continuous functions. One could disallow all continuous functions that are not polynomials, or quotients of polynomials, but that would yield algebraic geometry, not p-adic analytic geometry. On the other hand, allowing all analytic functions – convergent power series – would again not be geometric. Tate was the first to propose a workable compromise with his theory of *rigid analytic spaces*.

Tate's solution to the continuity problem was versatile, ingenious, and thoroughly in the spirit of Grothendieck's reformulation of topology. It also had a number of immediate applications to traditional problems in arithmetic geometry, notably the construction of the *Tate elliptic curve*, which still serves as a model for applications of Tate's rigid geometry. The field of *rigid analysis* that developed on the basis of Tate's (long-unpublished) founding document has its own canon and characteristic vocabulary; especially after it was extended by Michel Raynaud and Pierre Berthelot, it has grown into a healthy subfield of arithmetic geometry, a topic with its own practice and a resource for those in neighboring fields. However, it was not the last word; specifically, it lacked a native cohomology theory along the lines of étale cohomology, and the rigid analytic version of de Rham cohomology (*Monsky-Washnitzer cohomology*, later incorporated in Berthelot's *rigid cohomology*) lacked certain desirable features. This pointed to a failure of p-adic Hodge theory to be a true analogue of complex Hodge theory. Just as complex algebraic varieties are also complex analytic manifolds, p-adic algebraic varieties are also rigid analytic spaces. Hodge's Hodge theory is based on complex analysis, but the proofs of p-adic Hodge theory were purely algebraic. In particular, there was no p-adic analytic way to interpret the algebraic invariant of étale cohomology.

Three distinct versions of étale cohomology in the setting of p-adic analysis were proposed in the early 1990s, due respectively to Kazuhiro Fujiwara, Vladimir Berkovich, and Roland Huber. Each of these approaches involved a more or less radical expansion of the kinds of spaces considered in p-adic geometry, and each of them had immediate applications.[9] Huber's theory of *adic spaces* had relatively few adherents, however, until it was revived by Scholze as the setting in which perfectoid spaces naturally coexisted with other kinds of p-adic algebraic and analytic structures.

Before I introduce Scholze's perfectoid spaces, I need to devote a few words to the curious fact that algebraic geometry is so thoroughly algebraic that it can theoretically be developed without any reference to geometry whatsoever. The basic spaces in any version of algebraic geometry are the *affine* spaces. For Grothendieck, any commutative ring R defines an affine space – an affine *scheme* – denoted Spec(R), with the tautological property that R is the ring of functions on Spec(R). The correspondence between R and Spec(R) is an *equivalence of categories*, which means that everything you need to know about Spec(R) can be deduced from what you know about R, and vice versa.[10] Because Spec(R) is a kind of space, it

[9] This, by the way, is where I enter, and exit, the subject. The work discussed in chapter 9 of my *Mathematics without Apologies* was only possible because Berkovich had recently found a way to define l-adic cohomology for his analytic spaces. Fujiwara's theory played a crucial role in my subsequent work with Richard Taylor.

[10] However, this equivalence is *contravariant*; a comparison map (ring homomorphism) $R \to R'$ is equivalent to a comparison map of schemes Spec(R') $\to$ Spec(R).

has points, which are very convenient when you want to compare Spec(R) to Spec(R') for two rings R and R'. The general space in Grothendieck's version[11] of algebraic geometry is a *scheme* which is a topological space X that is covered by open sets that are *affine schemes*; in other words, each of whose points belongs to some Spec(R), or rather to a (generally infinite) collection of Spec(R'), compared in an appropriate way. One usually says that X has been constructed by *gluing* the affine schemes it contains. The geometric properties of schemes that can be reinterpreted as properties of their affine coverings – in other words, as properties of rings – are called *local*.

The prototype for constructing complicated spaces from simpler spaces is the theory of (topological or differentiable) *manifolds* in topology or differential geometry; the simpler spaces are just miniature versions of Euclidean space of dimension n, and a manifold is defined as a topological space that "looks locally" like Euclidean space. Differential geometry is to differential calculus as algebraic geometry is to ring theory.

In the various versions of p-adic geometry derived from Tate's rigid analytic geometry, spaces are constructed out of rings in the same way, except that the word *affine* is replaced by *affinoid*. Here is Scholze's definition of a perfectoid algebra (ring), copied from his article in the Proceedings of the 2014 International Congress of Mathematicians (ICM) at Seoul:

Definition 5.1. A perfectoid K-algebra is a Banach K-algebra R for which the subring $R^\circ \subset R$ of powerbounded elements is a bounded subring, and such that the Frobenius map $\Phi : R^\circ / p \to R^\circ / p$ is surjective.

Everything in the definition is important, but the most important part is the last line: it says that one can take p^nth roots for all n in this ring R°/p. To begin with, this allows Scholze to introduce the *tilt* of R, denoted $R^\flat$; the process of tilting turns a perfectoid algebra of characteristic zero into a perfectoid algebra of characteristic p in such a way that all the algebraic structures relevant to étale cohomology (among others) are preserved under the operation. The geometric objects Scholze attaches to the rings R and $R^\flat$ are adic spaces in Huber's sense; these are the *affinoid perfectoid spaces* that can then be *glued* together to form general perfectoid spaces.

The tilting operation originates in the early stages of p-adic Hodge theory, and specifically in a theorem of Fontaine and Jean-Pierre Wintenberger that identifies the Galois theory of certain infinite extensions of p-adic fields with the Galois theory of what are now called their tilts, which are fields of characteristic p. This operation is applied as the first step in constructing the

[11] There are still more general versions, but this will do for now.

most familiar of the Fontaine rings. Having learned from Grothendieck that any operation on a class of rings should be viewed as a *local* version of an operation on the corresponding class of spaces, arithmetic geometers naturally want to define a class of spaces to which the Fontaine-Wintenberger tilt extends that will for this reason be recognised as *the right* class of spaces. Several attempts were made, more or less rooted in the methods Faltings introduced in his work on p-adic Hodge theory; it is now acknowledged that Scholze's perfectoid concept is *the right* one for rings, and *the right* one for gluing the local pieces together into global geometric objects. The proofs made extensive use of the Faltings "almost ring theory."[12]

'Category' is the formalised mathematical concept that currently best captures what is understood by the word 'concept.' Scholze defined perfectoid spaces as a category of geometric spaces with all the expected trappings, and thus there is no reason to deny it the status of 'concept.' I will fight the temptation[13] to explain in any more detail just why Scholze's perfectoid concept was seen to be *the right* one as soon as he explained the proofs in the (symbolically charged) suburban setting of the IHES. But I do want to disabuse the reader of any hope that the revelation was as straightforward as a collective process of feeling the scales fall from our eyes. Scholze's lectures and expository writing are of a rare clarity, but they cannot conceal the fact that his proofs are extremely subtle and difficult. Perfectoid rings lack familiar finiteness properties – the term of art is that they are not *noetherian*.[14] This means that the unwary will be systematically led astray by the familiar intuitions of algebra. The most virtuosic pages in Scholze's papers generally involve finding ways to reduce constructions that appear to be hopelessly infinite to comprehensible (finite type) ring theory. This is my contribution to speculation about why Scholze succeeded so brilliantly where so many outstanding mathematicians failed.

THE CONCEPT'S RECEPTION

Scholze's first lectures on perfectoid spaces already included a stunning application, namely the proof of the weight-monodromy conjecture, in the

[12] As formalised and systematized by Gabber and Ramero (2003). Gabber, who is based at the IHES, attended Scholze's lectures, and I cannot resist mentioning how much arithmetic geometry is indebted to Gabber's constantly insightful questioning, which has kept the field honest for more than thirty years. At several points during his lectures, Scholze thanked Gabber for forcing him to clarify his ideas.

[13] I don't have to fight very hard, because I only barely understand the proofs myself. The interested reader can find the details in Scholze (2012).

[14] After Emmy Noether, the founder of modern algebra.

special case of complete intersections in projective space (extended in the published version to complete intersections in toric varieties) over p-adic fields. As was mentioned in the Prologue, this question was left unsolved by Grothendieck and Deligne. The latter had indeed proved the complete conjecture over the characteristic p analogues of p-adic fields – I can assure the reader that this was a very hard thing to do – and Scholze's methods allowed him to 'tilt up' Deligne's proof for the cases he was able to settle.

Six months before his IAS lectures, Scholze was seen as an exceptional talent by many colleagues, but not by everyone; a prestigious journal foolishly rejected one of his pre-perfectoid papers (which was later published in an even more prestigious journal). His proof of the weight-monodromy conjecture was enough to guarantee that this would not happen again. My understanding was that Scholze invented perfectoid spaces in order to 'tilt up' Deligne's proof of the weight-monodromy conjecture, but it was clear from the outset that the concept would have many more applications.[15] A few months after his IHES lectures, a French graduate student asked whether I would be willing to be his thesis advisor; things started conventionally enough, but very soon the student in question was bitten by the perfectoid bug and produced a Mémoire M2 – a mainly expository paper equivalent to a minor thesis – that was much too complicated for his helpless advisor.[16] By then Scholze had found two new spectacular applications that the precocious student managed to cram into his Mémoire M2, making it by far the longest Mémoire it has even been my pleasure to direct.

Although there have been more applications in the meantime – and I will mention at least one of them – I want to devote a moment to the two that were discovered in the year immediately following the IHES lectures, because they served to cement the idea that perfectoid spaces provide the right framework to adapt the constructions of complex Hodge theory in the p-adic setting. The first application was an (unexpectedly general) extension of the main theorems of p-adic Hodge theory to (proper) rigid-analytic varieties. This includes a new proof of the theorem of Faltings and Tsuji (and Nizioł) mentioned earlier; in its large structure and in its local arguments Scholze's is the proof that most closely resembles the proofs

[15] During the breaks in Scholze's lectures, there was active speculation about the applicability of his concept to one or another favorite problem.

[16] The student in question – who has been taken on by a second, more competent advisor – has not yet finished his thesis, but I would already count him as a member of the second perfectoid circle revolving around Scholze. The first circle, as I see it, includes Scholze's immediate collaborators and a few others; the second circle is already much broader, and there is a third circle consisting of everyone hoping to apply the concept to one thing or another.

of the main theorems of complex Hodge theory. The second application, joint with Jared Weinstein, was a classification of p-divisible groups over a complete algebraically closed p-adic field. All I want to say about p-divisible groups, is that they are objects of fundamental importance in arithmetic geometry and number theory, that they are distantly related to the Abel-Jacobi theory of period integrals on algebraic curves, and that there has been a dizzying wealth of classification schemes for p-divisible groups, each one useful for one purpose or another. Again, of all the classifications, the one of Scholze and Weinstein, apart from being perhaps the easiest to remember (though certainly not the easiest to prove), is the one that most closely resembles its complex counterpart. In both cases, the familiarity of the complex analogue, which is an advanced but necessary part of the training of practically any mathematician, reinforces the impression that Scholze has found *the right* way to think about p-adic geometry. And Scholze has himself contributed to this impression in his expository writing about perfectoid spaces.

Most of the blog posts, expository articles, and letters of reference I have seen agree that the most striking applications of perfectoid geometry are contained in Scholze's paper entitled "On torsion in the cohomology of locally symmetric spaces." A great deal can be said about this paper, which solved several outstanding conjectures in the course of vastly improving and generalising an earlier paper Harris, Lan, Taylor, and Thorne (2016) on a related topic. For the purposes of this essay, the importance of Scholze's paper is that it confirmed that perfectoid geometry provides *the right* framework for thinking about a number of central questions in algebraic number theory and, secondarily, that the cohomology of perfectoid spaces has *the right* (p-adic integrality) properties for applications to such questions.[17] And it confirmed the impression that number theorists had made *the right* decision to devote time to learning Scholze's new framework even before it had been shown to have important applications to their field.

DISCUSSION

I began writing this essay three years after Scholze's IHES lectures and one month after his ICM lecture in Seoul. One year earlier, I could safely assert that no one had (correctly) made use of the perfectoid concept except in

[17] The authors of the paper just cited – no doubt like many others working on similar questions – were actively seeking a p-adic cohomology theory with just these properties.

close collaboration with Scholze. The Seoul lecture made it clear that this was already no longer the case. Now, after nearly a year has passed, the perfectoid concept has been assimilated by the international community of arithmetic geometers and a growing group of number theorists, in applications to questions that its creator had never considered. It is an unqualified success.

How can this be explained? Mathematicians in fields different from mine are no better prepared than philosophers or historians to evaluate our standards of significance. One does occasionally hear dark warnings about disciplines dominated by cliques that expand their influence by favourably reviewing one another's papers, but by and large, when a field as established and prestigious as arithmetic geometry asserts unanimously that a young specialist is the best one to come along in decades, our colleagues in other fields defer to our judgement.

Doubts may linger nonetheless. I don't think that even a professional historian would see the point in questioning whether Scholze is exceedingly bright, but is his work really that important? How much of the fanfare around Scholze is objectively legitimate, how much an effect of Scholze's obvious brilliance and unusually appealing personality, and how much just an expression of the wish to have something to celebrate, the 'next big thing'? Is a professional historian even allowed to believe that (some) value judgements are objective, that the notion of *the right* concept is in any way coherent? How can we make sweeping claims on behalf of perfectoid geometry when historical methodology compels us to admit that even complex numbers may someday be seen as a dead end? "Too soon to tell," as Zhou En-Lai supposedly said when asked his opinion of the French revolution.

It is possible to talk sensibly about convergence without succumbing to the illusion of inevitability. In addition to the historical background sketched earlier in the chapter, and the active search for *the right* frameworks that many feel Scholze has provided, perfectoid geometry develops themes that were already in the air when Scholze began his career.[18] With respect to the active research programs that provide a field with its contours, it is understandable that practitioners can come to the conclusion that a new framework provides the clearest and most comprehensive unifying perspective available. When the value judgement is effectively unanimous, as it is in the

[18] These include the *Fargues-Fontaine curve*, a unifying construction due to Fontaine and Laurent Fargues, announced just over a year before Scholze's IAS lectures and generalised by Scholze, and work of K. Kedlaya and R. Liu on *relative p-adic Hodge theory*. Scholze described both of these in detail in his 2014 ICM talk.

case of perfectoid geometry, it deserves to be considered as objective as the existence of the field itself.

A value judgement applied to a new concept thus becomes an inflection or modality of a judgement of existence. I don't have much time for debates over the independent existence of the objects of mathematical practice, but I consider the existence of mathematical practice to be as objective as any other factual judgement. In that sense, I don't need to worry that the judgements by which my field's practice defines itself might be seen as tautological; what else could they be?

REFERENCES

Faltings, G. (1988). p-adic Hodge theory. *Journal of the American Mathematical Society*, 1(1), 255–299.

Gabber, O. & Ramero, L. (2003). *Almost ring theory*. Berlin: Springer Verlag.

Grothendieck, A. (1966). On the de Rham cohomology of algebraic varieties. *Publications mathématiques de l'IHÉS*, 29(1), 95–103.

Harris, M., Lan, K., Taylor, R. & Thorne, J. (2016). On the rigid cohomology of certain Shimura varieties. *Research in the Mathematical Sciences*, 3(1), #37.

McLarty, C. (2010). What does it take to prove Fermat's last theorem? Grothendieck and the logic of number theory. *The Bulletin of Symbolic Logic*, 16(3), 359–377.

Scholze, P. (2012). Perfectoid spaces. *Publications mathématiques de l'IHÉS*, *116*(1), 245–313.

PART V

9

Queering Mathematical Concepts

HEATHER MENDICK

WHAT ARE MATHEMATICAL CONCEPTS?

The proposal for this book opened with matters of definition. The authors noted that, 'Almost all research in mathematics education engages in one manner or another with mathematical concepts. The term concept, however, is often used without careful examination of its meaning. What exactly is a concept?' This orientation is not surprising given that most people working in mathematics education have backgrounds in mathematics. Indeed, those without a mathematical background generally lack credibility in the community. In mathematics, definitions are central to the discipline. From true definitions proceed further truths via logical rational argument, *or so we tend to assume*. In this way, we get the security and certainty in our knowledge that we desire, *or so we tend to assume*. In this chapter, I am writing deliberately against this desire to define.

I de-centre definitions by locating their truth not in their power to describe reality, but in their power to construct it. Truth is not something that precedes the act of defining and, for 'good' definitions, is embodied in it. Truth follows from this act, as definitions (re)shape the world to fit their purposes. My orienting question for this chapter then is not 'What exactly is a mathematical concept?' but 'What do mathematical concepts do?' This shift is a poststructural one, and I begin by outlining what that entails, moving on, in the second section, to explore how this can help us to look awry at mathematical concepts. Through this analysis, I argue that dominant ideas of the mathematical concept enforce rigid hierarchies of meaning with implications for how we understand mathematics and who can be successful at it. Thus, in the third section I offer reflections on 'What mathematical concepts could do?' if we queered them. Queer approaches define themselves, if they define themselves at all, by avoiding definitions, disrupting rigidities and subverting hierarchies.

My starting point for this chapter was my realisation upon reading the book proposal *both* that I do not know what a mathematical concept is *and*, after some anxiety about this, that it does not matter for the purposes of writing about it. A poststructural turn views mathematical concepts as enacted through discourses or collections of meanings:

> Discourses are structures of language and practice through which objects come into being. Discourses are knowledges about objects which are powerful because they determine what can be said about something as well as who can say it, and even what can be thought or imagined; they are, of course, imbued with values. (Mendick, 2006, p. 11)

We can see what difference a poststructural turn makes by applying it to an extract from G. H. Hardy's classic 1940 essay *A Mathematician's Apology*:

> A chair or a star is not the least like what it seems to be; the more we think of it, the fuzzier its outlines become in the haze of sensation which surrounds it; but '2' or '317' has nothing to do with sensation, and its properties stand out the more closely we scrutinize it … 317 is a prime, not because we think so, or because our minds are shaped in one way rather than another, but *because it is so*, because mathematical reality is built in that way. (Hardy, 1940/1969, p. 130; original emphasis)

Here, Hardy sets up a series of oppositions, the central one being between concepts like "2" or "317" which he includes within 'mathematical reality' and concepts like 'a chair or a star' which he inscribes into an unnamed but presumably lower and more mundane reality. Mathematical concepts are precise and beyond sensation, whereas other concepts are surrounded by a 'haze of sensation'. In identifying these oppositions, I am focused not on agreeing or disagreeing with Hardy's argument, but on how his argument works: what discourses and devices allow it to make sense. I might point to his use of quotation marks to bound his mathematical concepts, separating them from the fuzziness of the language that surrounds them; or to his use of italics to push his point home. Extending this outwards, I could embed Hardy's ideas in wider discourses of mathematical objectivity and rationality and point to the power that claims to objectivity and rationality carry in our society (Walkerdine, 1990).

'Discourse transmits and produces power; it reinforces it, but also undermines and exposes it, renders it fragile and makes it possible to thwart it' (Foucault, 1976, p. 101). Hardy acts as if he were telling us the way it is, but the very fact that he has to tell us, inadvertently, identifies mathematical reality as something that requires work to constitute it as *the way it is*. It is this work that discourses do that Foucault (1972) identifies when he speaks of them as "practices that systematically form the objects of which they speak" (p. 49).

For this chapter I have deliberately sought out examples of such discursive work. Borrowing from Sara Ahmed's (2010) idea of a 'happiness archive', I have compiled a mathematical-concept archive that I draw on across the chapter. Like Ahmed's, my archive is eclectic and structured through my encounters: "my archive is also my world, my life-world, my past as well as present, where the [idea mathematical concept] has echoed so powerfully" (Ahmed 2010, p. 19).

I encountered the first object I presented from this archive, *A Mathematician's Apology*, fifteen years ago while working towards my doctorate. It was an encounter that resonated with my earlier experience of reading about Hardy's friendship and collaboration with Indian mathematician Srinivasa Ramanujan. The 'rags-to-mathematical-riches tale' (*New York Times*, 1991; cited in Duchin, 2004, p. 5) of Ramanujan is an influential one: a self-taught poor Indian sends his ideas to a series of eminent mathematicians, and one among them, G. H. Hardy, perceives his genius and arranges for him to come to England and work with him. As Moon Duchin (2004, p. 5) points out, this is 'every bit the colonialist fairy-tale', and works to support the myth that we live in a meritocracy (Littler, 2013). In ignoring Ramanujan's elite position in Indian society and his access to education, it (re)produces discourses of 'natural genius'.

Crucial to the story we tell of Ramanujan is the way that he is said to have pursued his mathematical work through a combination of effort and intuition; he often claimed that his theorems derived from divine intervention. There is a curious synergy between mathematics as touching divinity and Hardy's idea of mathematics as, like God, beyond human endeavour. In this way, and as I explore further in this chapter, mathematics is frequently used to mark people out as special, as being out of, even above, the ordinary. Thus, discourses of mathematical concepts, like all discourses, are means through which people are positioned and position themselves. They "are infused by relations of power and practices of exclusion. Through them distinctions are made and certain relations, behaviours and people are given or denied value" (Allen & Mendick, 2012, p. 78). This begins to show how discourses of mathematical concepts relate to discourses about who can do mathematics. Tracing this relationship through further examples from my mathematical-concept archive is the focus of my next section.

LOOKING AWRY AT MATHEMATICAL CONCEPTS, OR WHAT DO MATHEMATICAL CONCEPTS DO?

Hardy's position feels extreme, positing a world of external truths to which only select individuals have privileged access. I opened with it because it is a familiar position in mathematics education, but one which it is tempting to dismiss as an easy target, as a position that few mathematics educators actually hold and so one with little influence over our field. However, my aim is to show why we should resist this temptation to dismiss it. For the discourses, or "fictions functioning in truth" (Walkerdine, 1990), that animate Hardy's statement remain the dominant ones through which we conceive mathematical concepts. I do this through close readings of two texts drawn from the apparently contrasting domains of popular culture and education policy.

Popular culture contains many representations of mathematicians and people doing mathematics, but few of mathematical concepts. US television crime drama series *Numb3rs* (Falacci & Heuton, 2005–2010) is a rare exception, and such concepts feature multiple times in each of its 119 episodes. This happens as the two central characters, brothers FBI agent Don Eppes and mathematics professor Charlie Eppes, team up to solve that week's crime, a process that requires the application of a range of mathematical concepts. These concepts are introduced through discrete interludes, usually under a minute long, that remove us from the programme's realist aesthetic and project us into a mathematical space. During these interludes, the conversational back and forth that dominates the rest of *Numb3rs* is replaced by a pedagogic voice-over by Charlie, or occasionally, in later seasons, by another of the mathematical characters. Visually, the context of the office or home is replaced by a mix of images, diagrams and symbols, set against a grid-like background. These objects move in patterns coordinated with the voice-over and other sounds. In short, we leave behind the fuzziness of everyday life and enter a mathematical space of order, clarity, beauty and structure.

In these interludes, mathematical concepts are conveyed through metaphors. For example, the Turing Test for artificial intelligence is explained through a discussion of real and artificial roses. There is neither sufficient time nor detail to allow casual viewers to access these concepts. Rather, the purpose of presenting them appears to be to provoke awe and wonder in the audience. The programme indicates that mathematics is useful, but that to access it we need an expert mediator like Charlie. As gatekeeper to mathematics, Charlie is marked as special in many ways in the series: he entered Princeton University at thirteen, published his first academic journal article at fourteen, holds a prestigious professorship, and manages to pursue original mathematical work alongside consulting for the FBI and writing a best-selling book about friendship. The strapline for the show may be 'we all use math everyday', but it is made clear in each episode that not all of us are able to use it to the same level. Thus mathematical concepts here act to mark out a few people as mathematical, and so special, and to exclude most from the possibility of acquiring those identities.

Turning to policy, the term 'concept' occurs fifteen times in the current English mathematics national curriculum, 'conceptual' occurs twice and '*mis*conception' once: concepts are both noun and adjective in mathematics and can be *mis*sed (Department for Education, 2014). Thirteen of the fifteen uses of the term 'concept' are attached to specific curriculum areas, including 'problem solving', 'addition and subtraction' and 'congruence and similarity'. While diverse, statistics and probability, a key strand of the curriculum, is markedly absent from this collection of concepts. This suggests how discourses of concept enact distinctions between areas of mathematics. Similarly, in three cases, these mentions of concepts are supplemented by distinctions between concepts and 'language' or 'vocabulary', for example, in the requirements that learners be able to 'use the concept and language of angles' or 'the concepts and vocabulary of prime numbers'. We also find distinctions between 'facts' and 'concepts'. These distinctions follow the same patterns as those in *A Mathematician's Apology* and *Numb3rs* where mathematical reality is distinguished from the fuzziness and mundaneness of the non-mathematical world. Concepts do work in these distinctions, holding the place of value over the lesser domains of facts and language, constructing a mathematics that can and should be abstracted from its 'context'. These are ways that value is distributed. They become particularly pernicious when applied to people.

The first of three aims for the curriculum introduces the idea of 'conceptual understanding', expressing the intention that children and young people:

> become fluent in the fundamentals of mathematics, including through varied and frequent practice with increasingly complex problems over time, so that pupils develop conceptual understanding and the ability to recall and apply knowledge rapidly and accurately.

'Conceptual understanding' recurs later in the document. The use of adjectives like 'conceptual' to qualify the type of understanding someone holds has become commonsense in mathematics education (Hossain, Mendick & Adler, 2013; Llewellyn, 2012). Notably, Liping Ma's (1999) comparative study of Chinese and US primary mathematics teachers argues that what marks the difference between them is not their knowledge, but the kind of understanding they have of mathematics: conceptual for Chinese teachers and procedural for US teachers. Alan Schoenfeld (quoted in Shulman, 1999, p. xii) describes Ma's book as "the only manuscript I know that has the attention and favor of both sides of the 'math wars'". He intends this as a compliment. However, a poststructural approach identifies as most problematic exactly those things that have become commonsense, things that seemingly oppositional positions hold in common, because they pass unquestioned.

Ma, in tune with others in mathematics education, does not present procedural and conceptual understanding as simply two contrasting approaches to the subject. She asserts conceptual understanding as superior and those who have it as better teachers of primary mathematics. Creating distinctions within distinctions, she opposes 'pseudo-conceptual understanding' to 'authentic conceptual understanding'. This suggests that there are real (aka authentic) and fake (aka pseudo) ways of being mathematical, and draws on the cultural value attached to authenticity, to 'keeping it real', to attach value to one and strip value from the other. These intricate distinctions are part of the work that mathematical concepts do; work as the 'highest' form of mathematical understanding is defined by its relation to concepts.

As this analysis demonstrates, distinctions are never innocent. Jacques Derrida (1981/2002, p. 39) writes, "[W]e are not dealing with the peaceful coexistence of a *vis-à-vis*, but rather with a violent hierarchy. One of the two terms governs the other (axiologically, logically, etc.), or has the upper hand" – the mathematical over the non-mathematical, the authentic over the pseudo, the conceptual over the procedural. I end this section by turning to the 'violent hierarchy' of gender (Wilchins, 2004) to show that the governing terms in these oppositions are not equally available to all. Similar arguments could be made in relation to hierarchies of 'race' or social class (see Stinson, 2013; Walkerdine, 1988).

Valerie Walkerdine, working with the Girls and Mathematics Unit in London during the 1980s, studied the gendering of mathematics from the early years to upper secondary education. The Unit had been set up to research girls' underachievement in mathematics. But what they found was that girls in primary school were *not* generally doing worse than boys in mathematics tests. However, even when girls achieved similar or higher marks to boys, their performance was taken to indicate their inferiority at mathematics. A decade before Ma, Walkerdine (1998) wrote:

> The principles used to explain performance in primary school Mathematics derive from theories of cognitive development: successful Mathematics performance is the attainment of concepts, particular stages of logical thought. In this view of learning, real understanding (based on concepts) is to be contrasted with rule-following or rote-memorization … which yields success without the secure foundation of real understanding. (p. 29)

The topics on which girls, on average, performed better were discursively associated with 'rote learning', while those where boys did better were constructed as requiring 'real understanding'. As Walkerdine identifies, concepts are at the heart of these distinctions: boys were attaining understanding based in concepts, girls were not. That this could become the truth about girls and mathematics against evidence of girls' test results should be remarkable. That it passed un-remarked before Walkerdine's work is because it aligns with long-standing Western discourses of gender, in which women are seen as handmaidens to the authentic creativity of men, in which women work hard but men, from Srinivasa Ramanujan to Charlie Eppes, are naturally able (Duchin, 2004; Mendick, 2006). These are the same discourses that forty years earlier allowed Walt Disney to define the "creative work' of animation as men's domain and to employ women at 'tracing the characters … and filling in the tracings … according to directions" (quoted in Noriega, 2015). They impact not just on how others position us but also on how we position ourselves, what each of us comes to imagine as possible and impossible for our lives. This is why we need to challenge dominant discourses of mathematical concepts.

QUEERING MATHEMATICAL CONCEPTS, OR WHAT COULD MATHEMATICAL CONCEPTS DO?

Thus far, I have identified some of the work that concept/ual does within mathematics: how it is implicated in detaching the subject from 'the world' and in setting up some people as 'naturally' successful at mathematics and others as 'naturally' unsuccessful. For the next stage of my argument, I draw a queer analogy between the mathematical concept and the Christmas season:

> The depressing thing about the Christmas season – isn't it? – is that it's the time when all the institutions are speaking with one voice. The

> Church says what the Church says. But the State says the same thing: … legal holidays, long school hiatus, special postage stamps and all. And the language of commerce more than chimes in, as consumer purchasing is organized ever more narrowly around the final weeks of the calendar year, … The media, in turn, fall in triumphally behind the Christmas phalanx. (Sedgwick, 1993/2002; original emphasis)

Eve Kosofsky Sedgwick's discussion of what is so depressing about Christmas also captures what is so depressing about mathematical concepts. The category of concept appears fluid, being approached from different philosophical positions and including and excluding different ideas in different times and places. Yet this fluidity is superficial, being subsumed within sameness, whereby the institutions speak with one voice. As argued earlier, whether in state curricula, pedagogic practice, initial teacher education, systems of assessment or media representations, concepts consistently do the same thing. Indeed, even people adopting radically different positions on the purposes and pedagogies of mathematics education take similar positions on concepts – their underlying significance to mathematics, their centrality to authentic mathematical understanding and the way that this marks out those who can from those who cannot. Taking inspiration from Sedgwick (2002/1993), I want to queer mathematical concepts, to disrupt the seemingly universal desire for the conceptual, to ask: "What if instead there were a practice of valuing the ways in which meanings and institutions can be at loose ends with each other? What if the richest junctures weren't the ones where *everything means the same thing*"?

Queer theory began with attempts to disrupt the fixed binary categories of gender (male/female) and sexuality (heterosexual/homosexual). It may seem strange (queer even) to be applying queer theory to something other than gender and sexuality, yet "a lot of the most exciting recent work around 'queer' spins the term outward along dimensions that can't be subsumed under gender and sexuality at all" (Sedgwick, 2002/1993). For example, J. K. Gibson-Graham applies queer theory to economics, seeking to disrupt the dominance of capitalism:

> What if we were to 'queer' mathematical reality and break apart some of its consolidating associations? We could start by reimagining the body of mathematics, that hard and masculine body that penetrates non-mathematics but is not itself susceptible to penetration … But how might we re-envision that body as more open and permeable, as having orifices through which non-mathematics might enter? (adapted by me from Gibson-Graham, 1996, p. 544)

I want to offer just one example of an unlikely orifice through which the 'non-mathematical' recently came to enter the 'hard … body' of mathematics: the 2015 EdExcel GCSE Mathematics examination. I do this in the hope that by letting go of the conceptual we can open up mathematics to other ways of being.

In England and Wales, usually at the age of sixteen, young people take GCSE (General Certificate of Secondary Education) examinations. There is some choice, but mathematics, English and science are compulsory. In most subjects, GCSEs are the culmination of two years of study, but for the compulsory subjects, studied since primary school, they are essentially a judgement on more than a decade of schooling. For most young people, their GCSE is their final mathematics examination. Working within government parameters, these examinations are set by a range of examination boards that compete for business. One of these boards is EdExcel, and in June 2015 teenagers used Twitter and other social media in unprecedented numbers to protest about that summer's examination.

While many young people took to the web to call for the grade boundaries to be lowered, most used humour to challenge EdExcel. On YouTube, they uploaded re-subtitled scenes of Hitler's anger and disappointment from the 2004 film *Downfall* to express their feelings towards the examination (for example, Campbell, 2015; momoneymopingu, 2015). The questions that provoked most comment in these and via #EdExcelMaths on Twitter were 'context questions' in which mathematical concepts were related to 'real life'. The main target was the now-infamous Hannah's sweets question:

> There are n sweets in a bag. 6 of the sweets are orange. The rest of the sweets are yellow. Hannah takes a random sweet from the bag. She eats the sweet. Hannah then takes at random another sweet from the bag. She eats the sweet. The probability that Hannah eats two orange sweets is 1/3. Show that $n^2 - n - 90 = 0$. (quoted in Bellos, 2015)

On YouTube, Hitler imagines telling Hannah where to 'stick her sweets' and to 'fuck off' with 'her shitty bastard sweets'. This aggression is also apparent in some tweets. Yet beyond this, reading through the lengthy Twitter stream, Hannah becomes a person with a family, friends, body and personality:

> Let's just hope that hannah or any other of her family member re-appears on this exam.
>
> I hope Hannah doesn't eat anymore orange sweets today. If she keeps that up she might have to go to the dentist.

> tomorrows maths paper better go rapidly uphill to make up for hannah & her inability to count how many sweets she has herself.
> What a ripoff? £15 for a revision day with 2 other schools? £15? Hannah could have eaten more sweets with that money.
> Calculate the area of Hannah's dead body after she choked on her sweets.
> So Sandra, Nalini, Sumeet, Steve and Fred have ruined my life. Along with Hannah, Karl, Mary and Jane.
> I'd like to personally thank Uditi [featured in another Edexcel sweets question] for not eating any sweets and being a greedy sod like Hannah.
> If one more equation shows up on this paper I am legit hunting down Hannah and making her explain to my parents why I failed.

The movement from probability to algebra will likely have made this question a particular target, as will its difference from the standard, much-practised questions on past papers. However, it is not the 'mathematical concepts' but the context of Hannah that attracted most attention online. Similarly, it is the other 'real life' contextual questions and the people in them who became targets, as in the tweet above that lists names of those who have 'ruined my life'. Some 'fe[lt] bad for Levi not having enough ribbon for his cake', Dave worrying about saving 2p on a sachet of coffee, 'Steve travelling from ashton to Barnsley at 1.7mph' or Karl raising only £10 for charity: 'Karl should have done a revision day and charged £15 per person, he'd have made an absolute tonne for charity'. Some young people made connections between them, for example, suggesting that Karl give his money to Levi so that he can purchase enough ribbon to go round his birthday cake.

As mathematics educators, we see Hannah and her sweets (and Levi and his ribbon and so on) as an irrelevance, quickly stripping her away in order to extract, in Hardy's terms, the mathematical reality of the question from the haze of sensation that surrounds it. Those protesting on Twitter persistently and insistently resisted this dominant mathematical imperative to dismiss the context as 'artificial' and cut through to the 'pure' mathematical concept. Returning to Hitler in his bunker, we find him upset that he cannot go to popular chicken restaurant chain Nandos without thinking about Kirsty and whether or not her chicken is the best value for money of those on offer. Instead of dismissing these young people as failing, lacking the necessary conceptual understanding and capacity to abstract, we can take a lead from them. Perhaps abstraction rather than connection is the problem. Could refusing to abstract and insisting on 'context' as inseparable from 'concept' have made any difference to the actions of those developing the collateralised debt obligations that precipitated the 2008 financial collapse?

Mathematics sits at the centre of political modernity, orienting us towards a future that is always already better than the present, even as we hurtle towards increasing economic inequality and environmental devastation. Thus, queering mathematics is more than an intellectual game; it is a way of enabling different political and pedagogical presents.

WHERE DO WE GO FROM HERE?

It is impossible to think completely outside the powerful commonsense that defines and is defined by the mathematical concept. For those of us who grew up with intimate relationships with mathematics, we are deeply invested in them, even if we now try to think against them. This became apparent to me recently while I was co-editing a book entitled *Debates in Mathematics Education.* An otherwise excellent chapter on teachers' use of examples troubled me because it lacked any 'debate'. In my feedback to the author, I suggested some ways that he could position the chapter's content in relation to debates about the form examples take and their relationship to the wider world. The author strongly rejected my suggestions, emailing:

> My chapter stands or falls on a simple proposition: since teachers DO choose and use examples all the time, it is better for their students if they choose them with care. It's about the actual example, not the context (or lack of context) in which it is set. I am NOT advocating a decontextualised, transmission pedagogy. Getting into a debate about the place of context in mathematics teaching would divert me (and the reader) from the intended purpose of the chapter.

I disagreed but accepted the chapter as it was. Later I realised that I too had embodied this distinction between mathematics and its context through the very structure I had set out for the book with an opening section on the 'context of mathematics education' creating the possibility for chapters in the sections on 'teaching and learning' and 'curriculum and assessment' to ignore context. As if context could be contained. As if the word 'context' could even begin to capture the implications of mathematics in the world and the world in mathematics. Perhaps this is why I have spent so many hours reflecting on the foregrounding of context and the stubborn refusal to abstract in young people's reactions to Hannah's sweets and #EdExcelMaths.

As a poststructural researcher in mathematics education, I have got used to the 'so what?' question. Reviewers and other readers of my work often ask what it means for practice, what difference it makes, what do we do, as if reading and thinking were not doing anything. Yet more than this, the rush to *do something* underestimates the power of the oppositional logics we are struggling with and against. As Derrida (2002/1981) writes, we must first overturn an opposition:

> To overlook this phase of overturning is to forget the conflictual and subordinating structure of the opposition. Therefore one might proceed too quickly to a *neutralization* that *in practice* would leave the previous field untouched, leaving one no hold on the previous opposition, thereby preventing any means of *intervening* in the field effectively. (p. 39; original emphasis)

Queer requires us to stay with our current confusion, and to resist new stabilities that promise to solve problems and make things better in the future. I thus end by listing (some of) the binaries I have been working to overturn in this chapter:

Mathematical reality vs Everyday reality
Numbers vs Words
Precision vs Haziness
Objective vs Subjective
Order vs Disorder
Mathematically able vs Mathematically unable
Special vs Ordinary
Mathematical concepts vs Mathematical language
Concepts vs Facts
Conceptual understanding vs Procedural understanding
Authentic conceptual understanding vs Pseudo-conceptual understanding
Real vs Fake
Real understanding vs Rote learning
Naturally able vs Hard working
Creative vs Imitative
Original vs Derivative
Masculine vs Feminine
Male vs Female
Concept vs Context
Algebra and probability vs Hannah and her sweets
The actual example vs The context in which it is set
Mathematical substance vs Mathematical diversion

I leave it to readers to explore – or not – ways of engaging with mathematical concepts so that these oppositional meanings, rather than lining up tightly, can be at loose ends with each other, rather than all saying the same thing, can begin to say something different.

ACKNOWLEDGEMENTS

I would like to thank: Tim Rowland for providing provocation and reassurance in relation to this chapter and allowing me to quote from an email correspondence with him; a great group of people at the BSRLM Day Conference in Dublin who engaged with my initial ideas; and four beginning mathematics teachers at Southfields Academy who helped me to think about Hannah's sweets.

REFERENCES

Ahmed, S. (2010). *The promise of happiness*. Durham, NC: Duke University Press.

Allen, K., & Mendick, H. (2012). Young people's uses of celebrity: Class, gender and 'improper' celebrity. *Discourse: Studies in the Cultural Politics of Education*, 34(1), 77–93.

Bellos, A. (2015). *How to solve the maths GCSE question about Hannah's sweets that went viral*. Retrieved 22 July 2015, from http://www.theguardian.com/science/alexs-adventures-in-numberland/2015/jun/05/how-to-solve-the-maths-gcse-question-about-hannahs-sweets-that-went-viral

Campbell, B. (2015). *Hitler reacts to Edexcel GCSE Maths 1MA0 Paper 1*. Retrieved 22 July 2015, from www.youtube.com/watch?v=QgTSe5InLbg

Department for Education (2014). *National curriculum in England: Mathematics programmes of study*. Retrieved 8 July 2015, from www.gov.uk/government/publications/national-curriculum-in-england-mathematics-programmes-of-study/national-curriculum-in-england-mathematics-programmes-of-study

Derrida, J. (1981/2002). *Positions*. London: Continuum.

Duchin, M. (2004). *The sexual politics of genius*. Retrieved 8 July 2015, from http://mduchin.math.tufts.edu/genius.pdf

Falacci, N. & Heuton, C. (Writers). (2005–2010). *Numb3rs*. New York: CBS.

Foucault, M. (1972). *The archaeology of knowledge*. London: Routledge.

(1976). *The history of sexuality (volume 1): The will to knowledge*. London: Penguin.

Gibson-Graham, J. K. (1996). Queer(y)ing capitalist organization. *Organization*, 3(4), 541–545.

Hardy, G. H. (1940/1969). *A mathematician's apology*. Cambridge: Cambridge University Press.

Hossain, S., Mendick, H. & Adler, J. (2013). Troubling "understanding mathematics in-depth": Its role in the identity work of student-teachers in England. *Educational Studies in Mathematics*, 84(1): 35–48.

Littler, J. (2013). Meritocracy as plutocracy: The marketising of 'equality' under neoliberalism. *New Formations*, 80/81, 52–72.

Llewellyn, A. (2012). Unpacking understanding: The (re)search for the Holy Grail of mathematics education. *Educational Studies in Mathematics*, 81(3), 385–399.

Ma, L. (1999). *Knowing and teaching elementary mathematics*. Mahwah, NJ: Lawrence Erlbaum.

Mendick, H. (2006). *Masculinities in mathematics*. Maidenhead, UK: Open University Press (McGraw-Hill Education).

momoneymopingu (2015). *Hitler reacts to GCSE Edexcel Maths (Paper 1 June 2015)*. Retrieved 22 July 2015, from www.youtube.com/watch?v=A5QfTEVw3ic

Noriega, M. (2015). *Read the rejection letters Disney used to send any woman who wanted to be an animator*. Retrieved 22 July 2015, from http://www.vox.com/2015/7/16/8978351/famous-rejection-letters-disney-animator-women

Sedgwick, E. K. (1993/2002). *What's queer?* Retrieved 22 July 2015, from http://users.clas.ufl.edu/kristinjoos/spring02/syp3000/queerreading.html

Shulman, L. S. (1999). Introduction, in Ma, L., *Knowing and teaching elementary mathematics* (pp. ix–xii). Mahwah, NJ: Lawrence Erlbaum Associates.

Stinson, D. (2013). Negotiating the "white male math myth": African American male students and success in school mathematics. *Journal for Research in Mathematics Education*, 44(1), 69–99.

Walkerdine, V. (1988). *The mastery of reason: Cognitive development and the production of rationality*. London: Routledge.

(1990). *Schoolgirl fictions*. London: Verso.

(1989/1998). *Counting girls out* (2nd ed.). London: Falmer.

Wilchins, R. (2004). *Queer theory, gender theory: An instant primer*. Los Angeles, CA: Alyson Books.

10

Mathematics Concepts in the News

RICHARD BARWELL AND YASMINE ABTAHI

Any answer to the question "What is a (mathematical) concept?" depends on who is asking, on who is answering and on the context in which the question is posed. This point holds in any situation but is, we argue, particularly visible in popular debates about mathematics education. The way in which the term "mathematical concept" is presented and discussed depends on the discursive, political context in which it is generated and used. To get an initial idea of what we mean, look at the following quotation, extracted from a news report published in Canada's *Globe and Mail* newspaper:

> A multi-pronged approach to modernize British Columbia's education system will see a shift toward concept-based learning … However, some critics are concerned the new approaches will overwhelm young minds and compromise foundational skills. …
>
> [Mrs H], a mother of two in North Saanich, says she has seen first-hand the confusion so-called discovery-based teaching techniques can lead to. "[My daughter] was being taught using Sudoku math puzzles [in Grade 3]. They had computer games in the classroom to learn the times tables," she said. "They were these methodologies to, I guess, conceptualize and make children think in different ways to come up with the answers. We don't have an issue with [that]. However, there wasn't a lot of emphasis on, say, learning the multiplication table."
>
> Her daughter continued to struggle with these concepts until she was enrolled in an after-school math program. "The transformation was incredible," [Mrs H] said. "After understanding their simple and effective methodology to solve math problems, it made her embrace math again. We went back to the basics."[1]

[1] Andrea Woo, "B.C.'s educational reforms are running into resistance", *The Globe and Mail*, 9 January 2014.

This news report explicitly discusses mathematical concepts and constructs a particular way of conceptualising mathematics. It links a new "concept-based" approach to learning to a process of modernisation of education in British Columbia. A mother is quoted who interprets this "concept-based learning", reframed as "discovery-based teaching", as focusing on encouraging children to "conceptualize and make children think in different ways". According to the report, these "concepts" were *confusing*, a situation that was only remedied by the introduction of a "simple and effective method" of solving mathematics problems. How can it be that learning mathematical concepts could be a problem for mathematics learning and teaching, while "simple" methods are not? To answer this question, we need to pose some others. How is the notion of "mathematical concept" constructed by this news report, and by news reporting about mathematics education in general? And what are these constructions of "mathematical concept" designed to do? In whose interests do these constructions operate?

In this chapter, we show how the construction of the term "mathematical concept" is shaped by discursive, political context. We draw on examples from Canadian news reporting on mathematics education to illustrate this idea. The use of examples from news reporting makes our point particularly clearly. Our argument, however, applies to any instance of the discursive construction of "mathematical concept", including in mathematics education research publications (see Barwell, 2009), in mathematics curriculum documents or in mathematics classroom interaction.

Alert readers may have noticed that we do not intend to give our own answer to the question, "What is a concept?" Rather, we are interested in how the term "concept" is constructed in, in our case, news reporting and shaped by political debates. In the next section, we discuss existing research on news reporting of mathematics education. We then present our theoretical and methodological approach, which is based on ideas from discursive psychology and media theory, showing how they apply to our interest in discourses relating to mathematical concepts. Finally, we look in more depth at some of the ways in which the concept of concept is constructed in a corpus of articles from Canadian print news media.

REPORTING MATHEMATICS EDUCATION

The popular press strongly influences public opinion and judgements about education (Eccles & Jacobs, 1986). Debates about mathematics education in particular can make for heated media headlines. The portrayal of mathematics education in the media partly mediates the relationship between public opinion and the mathematics education system – its teaching, learning, curriculum and so on. This reporting has, at times, become the focus of research, in order to understand how and why mathematics education is portrayed in the news and to see its effects on public opinion and other stakeholders (Eccles & Jacobs, 1986; Kee et al., 2012; Gruber, 2006; Waldow, 2009; Ertl, 2006; Pons, 2012). For example, Eccles and Jacobs' (1986) investigation of the impact of media coverage on parents' attitudes towards the role of gender in children's mathematics abilities showed that media coverage had a significant effect on the beliefs of parents who were exposed to it. Or, in their study, Kee et al. (2012) illustrated how the ruling political party in Malaysia was forced to react and respond to voices that were raised in news coverage regarding the decision to move to teaching mathematics and science in English.

In the United States, after the passage of the No Child Left Behind Act in 2001, issues related to assessment in mathematics education received widespread media coverage. Camara and Shaw (2012) review work that suggests that such coverage "tends to be biased and provides greater attention to the negative rather than positive side of stories, presents an incomplete or too simplistic view of educational issues, [and] exhibits a major lack of understanding of statistics and educational research" (p. 34). They also highlight how such coverage is often presented in terms of contestation or controversy, with two opposing sides set against each other.

This kind of negativity has been prominent specifically in relation to the publication of PISA results, which have generated extensive and controversial headlines in different countries. Pons (2012), discussing the reception of PISA in six European countries, referred to this media coverage as "PISA shock"; that is, the popular media focuses primarily on how each country is positioned in the PISA rankings, leading to "shocking" headlines. This coverage, in turn, leads to particular ways in which people and other stakeholders translate, analyse and respond to the results of PISA (Pons, 2012). For example, the publication of the first round of the PISA results in December 2001 had a "Tsunami-like impact" (Gruber, 2006, p. 195) in Germany, affecting educational policy-making discourse (Waldow, 2009) and fundamentally changing educational policy (Ertl, 2006).

Much of the aforementioned research examines the nature and impact of media coverage of mathematics education. Another strand of research has focused on the discourses produced by such coverage. Appelbaum (1995), for example, examined constructions of mathematics education in popular culture, including news media, in the United States in the 1980s. He argued that mathematics teaching was popularly constructed as a kind of heroic, individual endeavour. This construction makes it more difficult to think about mathematics education in social or political terms. The discourses about mathematics teaching identified in Appelbaum's (1995) research draw on discourses circulating in society, but also contribute to the reproduction of these discourses. More recently, Lange and Meaney (2014) noticed that a sample of media texts collected in Australia reflected a discourse of commodification of children in which children were seen as objects of investment (of time, of money, etc.) and to which "value needed to be added" (p. 392), again reflecting the link between mathematics education, rationality and a capitalist society.

In our ongoing research into Canadian news reporting of mathematics education (e.g., Barwell & Abtahi, 2015), we have noticed, in particular, a polarising discourse that constructs mathematics teaching and curriculum as a choice between two methods: "discovery learning" and "back to basics". This "choice" is apparent in the opening example, which also implies that discovery learning is confusing and therefore problematic, and that going "back to the basics" will remedy the situation. For this chapter, we are interested in how this polarising discourse shapes the construction of the concept of mathematical concept.

THEORETICAL BACKGROUND

The question "what is a (mathematical) concept?" implies some assumptions that, drawing on discursive psychology (Edwards, 1997; Edwards & Potter, 1992), we seek to avoid. The question assumes that (1) there is such a thing as a concept, and (2) with careful analysis or reasoning, we can identify, describe or define what a concept is. These assumptions, however, already presuppose a particular view of the world and, more specifically, a particular view of cognition. The term 'concept' is, after all, a cognitive one – in the *Oxford English Dictionary*, synonyms include 'idea', 'thought' or 'mental representation'. From the perspective of discursive psychology, however, it is not possible to directly examine, identify or describe ideas, thoughts or mental representations as they appear in human minds. One of the principal reasons for this position is an assumption that any attempt to access or describe things like concepts necessarily requires the use of language. But linguistic interaction between human beings is primarily a social process, in which any consideration of topics like concepts is necessarily situated within specific social contexts (Edwards, 1997; Edwards & Potter, 1992). In the *Globe and Mail* article at the start of this chapter, for instance, Mrs H's account of her daughter struggling with mathematical concepts is presumably selected from a longer interview and included in a newspaper article that presents a particular 'story'. Newspaper articles involve various requirements, including, most obviously, that they report "news" (and not simply "things that happened"). These requirements shape the way in which accounts like Mrs H's are presented and constructed and so cannot be read as an unvarnished account of experience. While many theoretical and methodological approaches to educational research recognise that language and social interaction are important aspects of human cognition, they do not tend to problematise the role of language and interaction in producing and constructing the data through which deductions about human cognition might be made.

Our position, then, precludes the possibility of answering the question "what is a mathematical concept?", at least ontologically. Discursive psychology offers an alternative approach. Social action, rather than cognition, is assumed to be the primary organising feature of human interaction. That is, human interaction is not primarily organised in order to transmit information about mental states from one person to another. It is not, therefore, possible to examine human interaction and read off information about the mental states of the participants (Edwards, 1997; Edwards & Potter, 1992). As teachers, for example, we know that when a student says "I get it", we cannot necessarily assume that they do. As analysts, however, we *can* examine how utterances about mental states are produced in their social and interaction context, in order to understand how they are organised and what, socially, they are designed to do. So "I get it" must be understood in its social and interactional context, perhaps as producing a request to be left alone, for example. Similarly, we can examine how the term "concept" is used in different situations in order to understand what it is used to do. This position has a series of implications (the following points are based on Edwards, 1991, pp. 517–518):

- References to "concepts" are always part of some text, utterance, argument, account, etc.
- Such references are therefore always implicated in some form of social action, such as reporting, criticising, blaming, justifying, undermining, etc.
- Situated references to "concepts" therefore perform moral work in relation to the participants who are producing or interpreting the accounts in question (because these social actions implicitly or

explicitly include moral evaluations: blamings imply someone at fault; justifications imply someone is right, etc.).

- References to "concepts" may therefore usefully be examined to understand the kind of discursive work for which they are rhetorically designed, rather than for what concepts might actually be.

These points apply to any discursive situation. In journal articles reporting research on children's learning of mathematics, for example, the cognitive process "learning" is constructed using particular devices, such as contrasts and the careful construction of "neutral" descriptions of children's mathematical activity (Barwell, 2009). The action orientation of discourse is, however, particularly apparent in news reporting.

Framing

The media theory term "framing" refers to common structures used in news reporting to make it attractive and easy to read. Framing refers to "modes of presentation that journalists and other communicators use to present information in a way that resonates with existing underlying schemas among their audience" (Scheufele & Tewksbury, 2007, p. 12). Framing is based on the assumption that how an issue is characterised in news reports can have an influence on how it is understood by audiences; that is, it is rhetorically designed. News production is concerned with how forces and groups in society try to shape public discourse about an issue by establishing predominant labels. These labels are often characterised in terms of pairs of concepts that readers come to see as connected (Price & Tewksbury, 1997), such as the description of politics in terms of left and right. Such framings are often politically loaded. The Canadian news reports we have collected about mathematics education, for example, refer to forces of action and reaction by parents, governments, and ministries of education organised around a framing that contrasts "discovery learning" with "back to basics".

More specifically, framing (as with any discursive account) involves selection and salience (Entman, 1993). The central effect of a frame is created by the underlying interpretive schemas that have been made applicable to the issue. One way to frame news is morality framing, in which news is presented in human interest terms. Morality framing personalises or emotionalises the news, in order to capture and retain audience interest (Semetko & Valkenburg, 2000). Morality framing usually requires motivated human actors (e.g., corporations, government agencies) who bring about harm and whose acts cause people to suffer undeserved hardship. These frames are often built into narrative forms that dominate news reporting. The task of every frame is to interpret "the source of these sufferings and consider how they can be mitigated or eliminated" (Gamson, 1977, p. 36). What makes morality framing powerful is its strong relationship between the prominence of moral frames in media discourse and in popular discourse. That is, in a framing in which moral evaluation "is central and highly visible in media discourse, it is equally central and visible in the attempts of working people to make sense of the issue" (Gamson, 1977, p. 58).

News reporting about mathematics education is often heavily moralised, through, for example, quoting frustrated parents, struggling teachers or angry mathematics professors. Hence the ways in which the news constructs mathematics education in turn is likely to influence the general public and policy makers through the perpetuation of particular discourses about mathematics learning and teaching. In the following section, we look in more depth at some of the ways in which mathematics education in general, and then mathematical concepts in particular, are constructed in a corpus of articles from Canadian print news media.

CONSTRUCTING CONCEPTS

Mathematics education often makes the headlines. We looked at a corpus of fifty-three news articles about mathematics education, published in a six-month period (September 2013–March 2014) in three national news publications in Canada. The publications were the *Globe and Mail*, the *National Post* and *Maclean's* magazine. About half of the articles consist of coverage of recently released PISA mathematics results. In our analysis of this corpus, which we only summarise here, we looked at the construction of mathematics education and how it related to and produced broader political narratives on mathematics education. The main framing that emerged was based on the polarisation of mathematics education into discovery learning and back to basics. In this section, we discuss three extracts from the corpus, selected because they make particularly visible the way in which the concept of concept is constructed in the news. For each extract, we first show how it reflects the overall framing that we saw in the full corpus. We then examine how the notion of mathematical concept is shaped by this framing within the particular extract.

Our first extract comes from a news report published in the *Globe and Mail* newspaper:

> Parents in provinces such as Alberta, Saskatchewan and Ontario look at their children's math homework and see little, if anything, of the fundamentals they were taught just decades ago. Gone are the days, in much of the country, of long division, mad-minute multiplication, addition with a carry and subtraction with a borrow.
>
> Today, children in provinces that have introduced the Western and Northern Canadian Protocol (WNCP) curriculum – a vast swath of the country – learn instead by investigating ideas through problem-solving, pattern discovery and open-ended exploration …
>
> "What we're seeing is the final demonstration that things have been going downhill," Prof. Craigen said.[2]

In this extract, a contrast is established between "fundamentals", which were apparently taught in the past, and ("instead") methods referred to as "investigating ideas through problem-solving, pattern discovery and open-ended exploration". This contrast is created in the text – it is possible to imagine combining problem-solving, pattern discovery or exploration with establishing and using methods of long division or subtraction. The contrast between the two approaches, however, reflects the broad frame apparent in many of the articles we examined. The "fundamentals" represent the "back to basics" pole, while the alternative methods represent what we have termed "discovery learning". The contrast is reinforced by the surrounding language, including the implicit sense that things were better "in the old days", and that "a vast swath of the country" is being somehow taken over. It is more difficult to imagine the expression "vast swath" being used to describe something with which the author was in favour.

So what does the "discovery learning" vs "back to basics" frame accomplish? According to the theory of framing, the construction of two poles makes the topic easier for readers to follow and the reporting easier to interpret. It also makes possible a politicisation of mathematics education, as the extract shows. Without the two possibilities, and the narrative of "fundamentals" replacing "problem-solving, pattern discovery and open-ended exploration", the idea that "things" are going downhill makes little sense. That is, by creating the two clear alternatives, one of the two can then be positioned as desirable and the other as problematic. In particular, this contrast is related to political debates over the WNCP curriculum. The new curriculum is portrayed as the cause of children's (apparently) poor mathematics performance.

[2] Kathryn Blaze Carlson, "Math wars: The division over how to improve test scores", *The Globe and Mail*, 10 January 2014.

This overall framing shapes the way "mathematical concepts" are constructed. This point is implicit even in the above extract. The "fundamentals" are references to specific algorithms – procedures used to carry out division, subtraction, addition or multiplication calculations using pencil and paper. Indeed, the author has carefully made reference to all four operations, and the list suggests that there are single standard methods for each one. Concepts, however, are hinted at in the alternative list of activities, which refers to "investigating ideas". Thus the contrast between "discovery learning" and "back to basics" appears to place mathematical concepts with "discovery learning". Further evidence for this possibility comes from a later part of the article, in which provincial curricula are summarised and discussed. The following extract comes from a paragraph devoted to British Columbia:

> The ministry of education dictates what must be taught, but not how to teach it. Here's an example: A Grade 4 student is required to learn mental math strategies for adding two-digit numbers, but the teacher chooses whether to teach the strategies by adding from left to right, or top-to-bottom with a carry. Some critics argue the current math curriculum is too conceptually based, with parents in the province launching a petition calling for greater emphasis on basic math skills.[3]

The extract first gives a summary of one point from the mathematics curriculum, highlighting possible calculation strategies a teacher might teach. A criticism is then reported that the curriculum is "too conceptually based" and that some parents would like a curriculum that emphasises "basic math skills". Thus skills are contrasted with concepts, a contrast that appears to be derived from the general framing of "discovery learning" vs. "back to basics".

It is also worth noting that the construction of mathematical concepts within the broader news framing includes an implicit moral dimension. In the above example, frustrated parents are frequently cited as undergoing undeserved hardship, due to the discovery learning promulgated by the curriculum. This moral dimension is also apparent in the following extract:

> The problem is twofold. For one, straight long division isn't on the curriculum anymore; at least not as it once was. The old ways of learning – rote strategies and "math facts" – have been replaced by so-called "discovery math" and "inquiry-based" teaching methods that focus on

[3] Kathryn Blaze Carlson, "Math wars: The division over how to improve test scores", *The Globe and Mail*, 10 January 2014.

> word problems, strategies and estimations. Manitoba in 2008 formally adopted one of the most radical math curriculum overhauls, which turned out to be an abysmal failure and has since been scaled back. Other provinces have moved forward with discovery math programs under the Western and Northern Canadian Protocol (WNCP), which was launched in 1995 and updated in 2006. And parents, never mind children, are having trouble understanding the new concepts.[4]

The general framing is again apparent in the reference to "rote strategies and math facts" vs. "discovery math" and "inquiry-based teaching methods". This extract is also constructed around a narrative of decline, in which the "old ways", the "math facts" and "straight long division" have been replaced by "discovery math", "inquiry-based teaching methods" and "new concepts". This narrative is reinforced by reference elsewhere in the article to declining Canadian PISA scores. Calculation methods are again foregrounded, with "straight" long division mentioned again. The qualifier "straight" suggests a natural, unproblematic method, and is placed in contrast with so called "discovery math". In short, methods and facts are preferable to concepts and understanding.

The moral dimension is implied by the portrayal of parents and children as "having trouble understanding the new concepts". The "discovery learning" vs. "back to basics" framing thus constructs not only a simple choice between two approaches but also the idea that one of the choices ("discovery learning") is causing problems in the form of troubled parents and a national decline in mathematics performance. The alternative choice ("back to basics"), meanwhile, which has apparently been usurped by "discovery learning", was part of an earlier and better time, with the implication that a return to such an approach is necessary. Hence a preference for methods and facts over concepts and understanding is not justified in terms of students' learning, but in terms of the moral requirements to alleviate the difficulties faced by parents and the need to reverse a decline in test scores.

This moral problem also includes struggling teachers, angry mathematics professors and students who do not know what to do. For example, the article cited above also includes the following statement:

> a teacher without a firm grasp of mathematical principles, as well as the tools to explain them, shouldn't be the one explaining BEDMAS to the class. We need to get back to proven methods – back to basics – and leave the discovery to history and English.[5]

[4] Robyn Urback, "Math isn't hard. Teaching it is", *The National Post*, 5 December 2013.
[5] Robyn Urback, "Math isn't hard. Teaching it is", *The National Post*, 5 December 2013.

Explicit in this report is how, for example, the conceptualisation of BEDMAS (an acronym indicating what order of operations to follow in a compound arithmetic or algebraic expression: brackets, exponents, division, multiplication, subtraction) is constructed and shaped within the broader polarised and politicised portrayal of mathematics education. Much like long division, or "subtraction with a borrow", BEDMAS is arguably more procedural than conceptual. The choice of such 'concepts' is designed to fit within the overall argument in favour of "back to basics". The selection of such 'concepts' lends itself to talk of "proven methods", 'facts" and "fundamentals" in a way that alternatives may not (think of base 10 structure, for example).

We can now return to the extract shown at the start of the chapter and the question of how it could be that mathematical concepts are problematic. In the extract, "concept-based learning" is presented as potentially overwhelming for young minds and as potentially "compromis[ing] foundational skills". The notion of mathematical concepts as a focus of learning is clearly shaped by the general framing in terms of "discovery learning" vs. "back to basics". As with the other extracts we have discussed, a narrative is implied in which discovery learning has problematically displaced the basics. This essentially political framing has the effect of positioning the learning of mathematical concepts, which are aligned with discovery learning, against the learning of "foundational skills", aligned with back to basics. Moreover, learning mathematical concepts is constructed as "overwhelming", implying that learning foundational skills is not. Needless to say, there are alternative ways of constructing relationships between concepts and skills.

The account of Mrs H's experience has also been constructed within the general frame and in a way that is designed to support the preceding separation and the "problems" that arise from learning concepts in mathematics. So discovery learning caused confusion and is contrasted with "learning the multiplication table". The problem is remedied, in effect, by the application of "back to basics" in the form of the after-school math program's "simple" methods of solving mathematics problems. The point here is not to question the veracity of Mrs H's account, nor of the reporter's presentation of this account. Our point is rather that the presentation of Mrs H's experiences is clearly shaped by the broader frame, both in terms of the consistent contrasting of discovery learning with back to basics, the construction of the former as problematic and the latter as the solution and the construction of a moral dimension which gives urgency to the situation. Within this complex set of discursive forces, learning mathematical concepts become a source of confusion that can overwhelm students.

CONCLUSIONS

The terms "concept" or "mathematical concept" are useful, in the sense that they are part of various discourses about mathematics education (and other discourses) and as such enable people to discuss aspects of their research, or the curriculum or students' thinking, and so on. A discursive psychology perspective, however, rejects a view of language as a window on the mind. It is not, therefore, possible to know what exactly a concept is, or what a mathematical concept looks like in a student's mind, or what anyone really means when the use these terms. It is possible, though, to examine how these terms are used, in context, to achieve particular rhetorical or political purposes. In the examples we have discussed in this chapter, mathematical concepts have turned out to play a surprising role within the broader news discourse of mathematics education. Specifically, mathematical concepts are constructed in fairly negative terms. They are confusing, overwhelming and difficult to understand. And mathematics curricula that emphasise concepts or conceptual learning are positioned as having usurped previously successful, tried and tested approaches that focus on facts, skills and procedures. Such concept-based curricula are therefore implicated in declining test scores and, in some articles, with a general state of national economic decline.

In our corpus, then, mathematical concepts are shaped by the general frame that sets "discovery learning" against "back to basics". There is a political backdrop to this frame. Back to basics is a conservative position: things were better in the old days and we should return to these tried and tested methods. This is not to say that everyone who argues for such an approach has a conservative view of mathematics education; our point is about discourses, not individuals. The trouble with news frames is that it becomes difficult to discuss anything without reference to the frame. It is therefore difficult for alternative visions of mathematics education, and alternative constructions of mathematical concepts, to find a way into public discourses of mathematics education.

We have sought to show how the term "mathematical concept" is shaped by the local context in which it is used, as well as by broader political framings that set up a contest between two opposing positions. This general point applies to any situation in which terms such as "mathematical concept" are used – it is not simply an artefact of news reporting, although it is perhaps more striking in this context. Thus, the construction of "mathematical concepts" in educational policy and curricula is, we can speculate, shaped by the requirements of such documents to generate lists of carefully specified items that can be measured and assessed in some way, perhaps for the purposes of accountability. And indeed, our own presentation of this topic is shaped by the norms and requirements of research writing and relies on particular distinctions and contrasts, including, most notably, the distinction between discourses of "discovery learning" and of "back to basics". For us, then, the question of what a mathematical concept really is less interesting than the question of how the idea of mathematical concepts is deployed in the various different discourses of mathematics education.

REFERENCES

Appelbaum, P. (1995). *Popular culture, educational discourse, and mathematics*. Albany: SUNY Press.

Barwell, R. (2009). Researchers' descriptions and the construction of mathematical thinking. *Educational Studies in Mathematics*, 79(2), 255–269.

Barwell, R. & Abtahi, Y. (2015). Morality and news media representations of mathematics education. In S. Mukhopadhyay & B. Greer (Eds.), *Proceedings of the Eighth International Mathematics Education and Society Conference* (pp. 298–311). Portland, OR: Portland State University.

Camara, W. J. & Shaw, E. J. (2012). The media and educational testing: In pursuit of the truth or in pursuit of a good story? *Educational Measurement: Issues and Practice*, 31(2), 33–37.

Eccles, J. S. & Jacobs, J. E. (1986). Social forces shape math attitudes and performance. *Signs*, 11(2), 367–380.

Edwards, D. (1991). Categories are for talking: On the discursive and cognitive bases of categorization. *Theory & Psychology*, 1(4), 515–542.

(1997). *Discourse and cognition*. London: Sage.

Edwards, D. & Potter, J. (1992). *Discursive psychology*. London: Sage.

Entman, R. M. (1993). Framing: Toward clarification of a fractured paradigm. *Journal of Communication*, 43(4), 51–58.

Ertl, H. (2006). Educational standards and the changing discourse on education: The reception and consequences of the PISA study in Germany. *Oxford Review of Education*, 32(5), 619–634.

Gamson, W. A. (1977). *Talking politics*. Cambridge: Cambridge University Press.

Gruber, K.-H. (2006). The German 'PISA-Shock': Some aspects of the extraordinary impact of the OECD's PISA study on the German education system. In H. Ertl (Ed.), *Cross-national attraction in education: Accounts from England and Germany* (pp. 195–208). Oxford: Symposium Books.

Kee, C. P., Ibrahim, F., Ahmad, F. & Khiang, C. C. (2012). Frame contention between news sources and news media: Framing the dispute of teaching Mathematics and Science in English. *Asian Social Science*, 8(5), 16–47.

Lange, T. & Meaney, T. (2014). It's just as well kids don't vote: The positioning of children through public discourse around national testing. *Mathematics Education Research Journal*, 26, 377–397.

Pons, X. (2012). Going beyond the 'PISA shock' discourse: An analysis of the cognitive reception of PISA in six European countries, 2001–2008. *European Educational Research Journal*, 11, 206–226.

Price, V. & Tewksbury, D. (1997). News values and public opinion: A theoretical account of media priming and framing. In G. A. Barett & F. J. Boster (Eds.),

Progress in communication sciences: Advances in persuasion (pp. 173–212). Greenwich, CT: Ablex.
Scheufele, D. & Tewksbury, D. (2007). Framing, agenda setting, and priming: The evolution of three media effects models. *Journal of Communication*, 57(1), 9–20.
Semetko, H. A. & Valkenburg, P. M. (2000). Framing European politics: A content analysis of press and television news. *Journal of Communication*, 50(2), 93–109.
Waldow, F. (2009). Review essay: What PISA did and did not do: Germany after the 'PISA-shock'. *European Educational Research Journal*, 8(3), 476–483.

11

Concepts and Commodities in Mathematical Learning

TONY BROWN

Educational thought is undoubtedly ideological, but its application to mathematical ideas can seemingly anchor more radical ambitions. It is often thought that you are either right or wrong in mathematics, with little space between. Educational research either informs improvement or it does not. The advance of mathematics as an academic field more generally, however, is defined by the production of new ideas, or concepts, which adjust progressively to new ways of being. That is, mathematical concepts are created to meet the diverse demands of everyday life, and this very diversity can unsettle more standardised accounts. For example, the expansion of mathematics as a field often relies on research grants selected to support economic priorities. In schools, economic factors influence the topics chosen for a curriculum. Our evolving understandings of who we are and of what we do shape our use of mathematical concepts and thus our understandings of what they are. Moreover, public images of mathematics pull in a number of directions that produce alternative conceptions of the field of mathematics. These disparities of vision result in much variety in how mathematical concepts are materialised in everyday activity. They also point more fundamentally to the uncertain ontology of mathematics as a supposed field itself and its evolution according to the demands made of it. Yet more typically, mathematics as a field is thought to exist as a consequence of rationality or even as a matter of belief. Ideology, however, can shape notions of utility, rationality and belief. School mathematics, this chapter argues, has been reduced according to ideological schema to produce its conceptual apparatus, pedagogical forms and supposed practical applications (Lundin, 2012). It has been transformed as a result of ever more pervasive formal assessment demands in schools linked to the regulation of citizens, as part of what Althusser calls the *ideological state apparatus*. The chapter offers some important insights into how Žižek's work extends Althusser's model of ideology as applied to revising our understanding of mathematics itself. It asks how we think about mathematics through ideological lenses, and contemplates the forms that it takes in the hundreds of hours that it occupies in most people's school education. The subjective experience of those hundreds of hours may exceed the ideological parameters whilst remaining in the service of those ideologies by making us believe them through the sheer force of habitual action.

For many people, mathematics stands apart from everyday life. It is other (slightly odd) people who do the more complicated versions of mathematics. Žižek's multiple references to Stephen Hawking point to a widely held attitude to mathematics and provide a somewhat unsympathetic reading of Hawking's popularity:

> Would his ruminations about the fate of the universe, his endeavour to "read the mind of God", remain so attractive to the public if it were not for the fact that they emanate from the crippled, paralysed body, communicating with the world only through the feeble movement of one finger and speaking with a machine generated impersonal voice? (Žižek, 1997, p. 173).

Žižek contends that Hawking's iconic status "tells us something about the general state of subjectivity today" (ibid), encapsulated, perhaps, in e-connected individuals expressing themselves, or accessing the world, through technological apparatus. Specifically, Žižek sees Hawking's mode of performance positioning mathematics and physics beyond the reach of a broader public through Hawking being a "new type of public intellectual … who, in the eyes of the wider public, stands more and more for the one 'supposed to know'[1], trusted to reveal the keys to the great secrets which concern us all" (Žižek, 2001, p. 212). Here mathematics resides in a parallel universe available only to those able or prepared to temporarily sacrifice everyday life to pursue the beauties of more abstract thought, and in so doing downplay analytical opportunities that could be more widely available through mathematical thought understood in a more inclusive way.

For various reasons, however, many people decline the benefits of a mathematical education. A recent report in Britain claimed that only 50% of adults function above the level of an average eleven-year-old, and very

[1] "[A] displacement of our most intimate feelings and attitudes onto some figure of the Other is at the very core of Lacan's notion of the big Other; it can affect not only feelings but also beliefs and knowledge – the Other can also believe and know for me. In order to designate this displacement of the subject's knowledge onto another, Lacan coined the notion of the *subject supposed to know*." (Žižek, 2006, pp. 27-29)

often members of the other 50% were quite proud of their limitation (Paton, 2012). The report's author was rather concerned: "Now that's a scary figure because it means they often can't understand their pay slip, they often can't calculate or give change, they have problems with timetables, they certainly can have problems with tax and even with interpreting graphs, charts and meters that are necessary for their jobs" (ibid, para 9). In addressing this problem the wonder of mathematics takes on a very different style in many school contexts that are shaped by teachers' accountability to examination regimes designed to support the many and various mechanical processes or economic structures that govern our lives. The pedagogical or practical mediation pertaining to such regimes reshapes and *commodifies* mathematical concepts into objects that can be more readily tested or applied within these regimes. That is, for many students in schools the space of mathematics is marked out by mechanical skills and procedures supportive of an ideological agenda. These aspects are privileged over developing more intuitive powers or other aspects of mathematics. The pedagogical objects of school mathematics (multiplication tables, Pythagoras theorem, decomposition method of subtraction), however, still mark concepts that retain their structural place within mathematical thought that exceeds these ideological parameters, ways of mathematical thinking that are suggested beyond the bare symbols (e.g. conceptualising iteration to infinity, the sense of a rotation in an angle measure). These latter aspects of mathematics are "exempted from the effects of wear and tear [where the supposed field of mathematics itself] is always sustained by the guarantee of some symbolic authority" (Žižek, 1989, p. 18). We forgive mathematics all of its awkwardness in everyday life as we sustain a faith in something more pristine.

So, although the very existence of mathematics is linked to our practical applications, there is also some implied claim to an underlying truth in a more abstract sense. Recent research in mathematics education has pointed to how the existence of mathematics is underwritten by its materialisation in structures and processes (Palmer, 2011). Karen Barad (2007) has shown us that it is never entirely clear where the human stops and where the operation of cultural machinery begins. For instance, is the mathematics that Hawking generates in his mind or in his computer? It is this sort of dilemma that has fuelled mathematics education research in recent decades. Research in the area had often in the past been governed by Piagetian conceptions of the mind (Piaget, 1952). Children were seen as passing through successive developmental stages where it was the teacher's job to enable the children to reconstruct ideas as they followed the inevitable or "natural" route to maturity. Mathematics and the mind were seen as two separate entities that got to know each other in the classroom. The international conference on the Psychology of Mathematics Education has provided a long-term centre of gravity for international researchers in the more generic field of mathematics education. In the last couple of decades, however, discursive constructions have become more familiar. In these later models the focus is not so much on minds developing as on changing the story or structure that individuals follow. On the one hand in this scenario students can construct their own accounts of mathematics, bringing new sorts of mathematics into being to meet the needs of their personal circumstances. In some contemporary understandings of mathematical learning pupils are seen as investigating mathematics towards introducing their own individualised structuring of the landscape being encountered, further blurring the line between the individual human and the mathematical concepts that she produces (e.g. Brown, 2011). Conversely, on the other hand, policy makers can legislate particular versions of mathematics, a more centralised script as it were, and police their implementation towards greater conformity (Brown & McNamara, 2011).

The next section provides an account of how mathematical concepts are produced. This formation is then considered in the context of school mathematics centred on the generation of ideological constructs that come to house these objects. Examples are provided of how we might understand policy makers, teachers and students variously identifying with or being alienated from mathematical phenomena. The chapter concludes by theorising how individuals begin to believe the ideologies that govern their actions.

THE PRODUCTION OF MATHEMATICS

Žižek's associate, Alain Badiou, follows philosophers such as Bachelard, Lakatos, and Althusser in seeing science as a *practice* marked by the production of *new* objects of knowledge (Feltham, 2008, pp. 20–21), in much the same manner as Deleuze and Guattari see philosophy as "the art of forming, inventing, and fabricating concepts" (Deleuze & Guattari, 1996, p. 2). Mathematics as a field can be seen as evolving through reaching new generalisations in newly encountered conditions. Over a longer term, the absorption of mathematics into life results in the field of mathematics itself changing. Certain elements of mathematics have been touched more frequently by the need to support applications (e.g. statistical analysis of demographic trends). The field of mathematics itself has been marked out according to how it has been seen as supporting practical agenda. Some aspects are much more popular than others for this reason and tend to be more likely used in everyday life, or secure research grants, etc. In a recent BBC radio feature, a professor of mathematics challenged a director of a government research grants agency by claiming that one could only get research grants for statistics in the current climate, such is the drive of supposed applications. Accordingly, mathematics itself has been reconstituted to meet evolving social priorities and criteria. The historical circumstances that originally generated mathematical objects are often lost. The objects may have become a part of who we are such that we are no longer able to see them. Geometrical constructions, for example, are routinely built into our physical landscape such that we do not notice them any more. We become accustomed to moving around such landscapes and those ways of moving become part of who we are. For instance, circles are common entities and they have been featured in many of the stories that we have told about our world. We may feel that we have gotten to know circles from a lot of perspectives, which results in them acquiring a broad set of qualitative features. We use *circle* as a concept in building our world, and as a result circles become materialised or absorbed in the very fabric of our physical and conceptual world. Stellated octahedrons, in contrast, have been denied that level of intimacy and familiarity with humans (Figure 11.1). Geometrically speaking, there is no reason as to why one might be privileged over the other. Circles have been reified not because of any essential difference between them and, say, stellated octahedrons, but because of merely historical and political reasons. It is actually quite difficult to sort mathematical concepts according to which ones are empirically referenced like circles and those that are not so common in appearance or utility, such as stellated octahedrons.

Mathematics exists as models of knowledge that sometimes support empirical enterprises, but ultimately, as empirical support, the models always reach their limits. We can never use words to precisely specify what mathematics is as such. Yet this realisation does not assist us much with understanding the predictive capabilities of mathematics, which have real psychic effects in more abstract mathematical analysis, and material effects in practical enterprises such as building bridges, the effective analysis of economic models, everyday finance, etc. There is something more significant to mathematical conceptualisation that *needs* to be accounted for. It has a precision and produces results unlike other languages. Mathematics can guide us or structure our thinking, but it does not fix our ways of making sense. Mathematics introduces polarities around which discourse can flow and which result in actual impacts on the physical and social world. Yet the

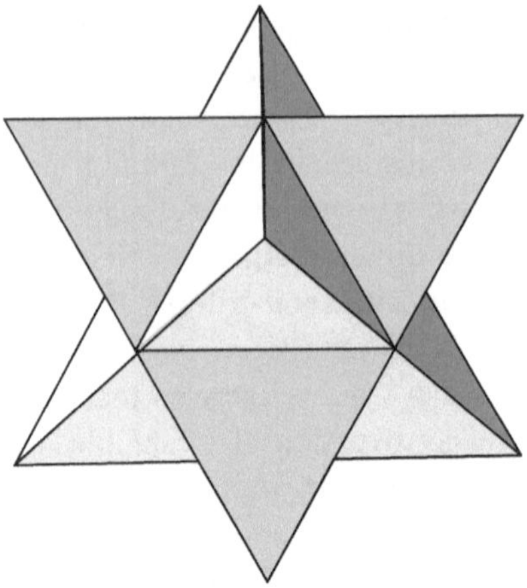

FIGURE 11.1 Stellated Octahedron.

possibility of mathematics as a complete system that supposes grounding in some empirical reference always slips away.

THE PRODUCTION OF SCHOOL MATHEMATICAL CONCEPTS

In the mundane of everyday life, children in schools learn or create mathematical procedures or pedagogical forms for handling different problems, such as those found in teaching schemes, textbooks or curriculums (e.g. how to teach the multiplication of fractions). The mathematics curriculum defines the forms through which school mathematical concepts are understood. For example, concepts of spatial awareness are learned through constructing triangles, reflecting shapes on graph paper, etc. Fashions change, learning theories move on, teaching schemes get replaced, resulting in school mathematics receiving regular makeovers whether or not these effect substantive changes (Brown, 2012). School mathematics has a tendency to reify particular objects (e.g. circles, the first ten integers, the formula for factorising quadratics) or procedures (e.g. the decomposition method of subtraction) for greater scrutiny. It is often applied mathematics shaped around recognisable situations. Particular configurations are repeatedly used resulting in the landscape of mathematics being viewed through perspectives that begin to characterise our engagement with mathematics. Questions are asked in familiar ways. Particular areas of mathematics are favoured, such as the concepts that are more easily tested (finding the difference between two integers, finding the area of a triangle) rather than exploring a mathematical terrain. Further, mathematical language used in schools points to styles of social interpretation, social practices and ways of understanding the teacher-pupil relationship.

In Žižek's (1989, pp. 11–48) notion of commodity-fetish, commodities (or specific forms) become the supposed objects of desire. In our case in question, *commodified* versions of mathematics have become the institutionalised markers, or concepts, of school mathematics (Brown & McNamara, 2011). The commodified objects, or mathematics' greatest hits, orientate our understanding of the subject. In Badiouian terms (Badiou, 2005), the objects of school mathematics result from certain sets of elements being "counted as one". For example, if mathematics is seen as needing to include learning multiplication tables, the emphasis on mathematical tables becomes part of the commodification of mathematics and the way it is understood more broadly. That is, the table compiling multiplication results becomes an object, a counting as one of a certain class of results that provide points of reference orienting the pupils' wider engagement with mathematics. The addition of elements to the school curriculum (e.g. tables and graphs) and the reduction of other areas (e.g. geometry) marks the ongoing historical formation of mathematics in the context of social practices. But the statements or concepts that locate mathematical phenomena so often become the statements that police its boundaries. Pupils must then know their tables if they are to advance in mathematics as it is understood within the particular regime.

Whilst university mathematics provides a system against which the correctness of school mathematics is judged, the latter is more often locally defined around social practices, such as calculating supermarket bills, estimating the number of bricks needed for a wall, predicting economic trends, etc. But why have classroom activities assumed the social forms that they have? That is, why have they become commodities with a given form? Commodification suggests a form of packaging designed for presentation in a particular way of life where worlds have been conceptualised around them. For Badiou (2009), the assertion of any object "counted as one" is linked to the assertion of a transcendental world. This conceptualisation of alternative worlds built around commodities mediates the production of mathematical concepts and proliferates or typifies the senses in which they can be understood. The core mathematical idea may be linked to a way of life, but in so doing it normalises particular forms of life as though they were a transparent layer free of ideology.

For example, the English mathematics curriculum has formatted mathematics for consumption in schools (Skovsmose, 1994). The government has exercised its control over teachers and students by specifying specific skills, conceptual awareness and competencies, which stand in for the government's supposed obligation to promote a numerate population with consequent benefits to our society, technology and the economy. Mathematics is characterised by the identification of a particular set of elements, which in turn imply a specific understanding of the world and how it might be changed. The route through which this can be achieved, however, may be difficult to specify in advance or interpret in retrospect. We may ask: what was in successive government ministers' minds in introducing such policy instruments into English schools (cf. Žižek, 2001, pp. 61–62)?

- The minister wanted to improve mathematics by whatever means as part of his quest to provide an education as a basic human right – any rationalisation of how he achieves this is secondary to that basic desire.
- The minister saw pursuit of the improvement of mathematics as a good ploy for re-election – his only real concern.
- The minister sincerely believed that the implementation of his policies will bring about improvement in mathematics in the way he suggests.
- The minister was himself aware that policy setting is not an exact science but instinctively believes that a simple and insistent presentation of his policies will achieve for him the best possible outcomes in some way or other. This might be through good participation among teachers, quantifiable improvements in test scores, an image of a government taking charge or, more negatively, the demotion of mathematics as a political issue in the public's eye.

Which account best describes the minister's perspective? Perhaps all of them do. It seems impossible to attain a "real" version of events governed by straightforward causal relationships. The options above merely provide alternative fantasies through which reality might be structured. To personify the implementation of policies with a clear association between one person's rational action and its effect risks oversimplifying the broader concern. The effects of policy implementation are probably too complex to be encapsulated instrumentally.

From a teacher's point of view one might contemplate reducing the emphasis on singular *metaphorical* associations between mathematical activities and mathematical concepts, in favour of a *metonymic* association between mathematical activity and social activity more generally. This entails linking the mathematical activities (seen as activities governed by certain procedures, rules, performance criteria, etc.) with other social discourses, including others specifically related to mathematics. The meaning of the mathematical discourses thus becomes a function of their relationship with the other discourses with which they are entwined, interpretive links that can always be revisited or renewed. This softens any assumption that the activities are anchored in specific mathematical concepts. Rather, we need to attend to the reification of such supposed concepts as they unfold in specific discursive environments. This would move us away from any supposed universal conceptions of what mathematics should be about; instead, it alerts us to the historical and social processes that generated classroom mathematics in the forms it now takes.

From a student's perspective, mathematics can often be presented as though it comprises singular answers to any given question, as if there is always a right and a wrong answer. This view of mathematics promotes a pedagogical attitude governed by the commodification of objects characterised by *this procedure getting that result*, verifiable rather than true. Yet it is possible to produce mathematics as a conceptually defined space in different ways. In some of my own teaching I designed some activities towards enabling the students to develop their spatial awareness as a prelude to a more formalised approach to geometry. They were invited to explore various body movement activities. In one such activity a student was asked to position herself between two fixed points on the ground that were about four metres apart such that she was positioned twice as far from one point as she was from the other. She was challenged to walk so that she was always twice as far from one point as she was from the other. A group of students observing this provided alternative interpretations of what was going on. A number of students produced drawings (Figure 11.2). Others provided algebra. Another reported on the emotional stress she experienced from being asked to do mathematics in this group situation.

The issues became more complicated as the problem shifted to remaining twice as far from one dot as from the other in three-dimensional space. The challenge provoked much gesturing alluding to points beyond immediate grasp. One student documented the different ways in which she saw her colleagues making sense of the problem:

> [P]eople do not visualise the same problem in the same way. When we were describing the same visualisation (e.g. the shape of the curve in 3D) each individual gave very different, but equally valid, explanations. For example, the explanations for seeing a circle in 3D were given as: a penny being spun around at the end of a piece of string; modelling the shape with your hands; imagining being the origin of the circle (therefore being inside the shape) and what it would look like looking in each direction; imagining the shape being built up from the established points which were on the ground.

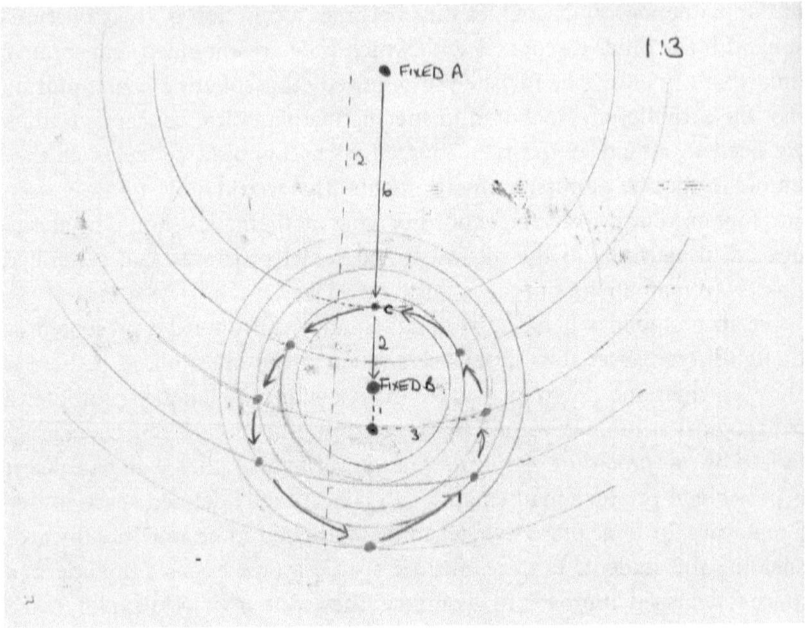

FIGURE 11.2 Circle.

Through such participation the activity became centred on documenting connections to alternative discursive formations of self: a physical self moving in space; a pedagogical self reflecting on the learning of others; a geometric self creating drawings; an algebraic self solving formulae; an emotional self on how it felt relating to other students. But in understanding oneself one is alerted to territory that one can grasp within the terminologies available and also spaces beyond reach that can only be pointed to or imagined from different perspectives. What had been movements of the body became materialisations of one's comprehension of reality itself. The experience of the configurations became linked to how one felt at the time, a narrative of participation formalised for posterity, for the time being.

THE INCOMPLETE PRODUCTION OF MATHEMATICAL REALITY THROUGH COMMODIFICATION

Can we then be more precise in depicting how mathematical concepts intervene in more ideological constructions of reality, where forms of practice motivate specific understandings of mathematical concepts? In a famous debate Richard Dawkins represented a rationalist camp that "raged against any kind of mystery in the cosmos, preferring instead to settle for a cold universe driven by the machine of pessimistic reason" (Žižek & Millbank, 2009, p. 6). He was countering Alister McGrath, a professor of theology, who had posited a religious thinker governed by faith. A second debate, however, between the theologian John Millbank and Žižek led to an assertion that faith and reason are not simply opposed to each other (ibid). They each argued in different ways that the work of Hegel undermined any dichotomy between the mythical and the rational. For Hegel, "the object [or concept] is always-already bound up in the complex mediating process of the subject's thinking it, and conversely, the subject's thinking the object is itself bound up in the object's very existence" (op cit., p. 14). "What we experience as reality is not the thing itself, it is always-already symbolised, constituted, structured by way of symbolic mechanisms" (Žižek, 2011, p. 240). Its very constitution is ideological. Žižek (2011, p. 144) identifies three positions in Hegel's formulation:

> In the first, reality is simply perceived as existing out there, and the task of philosophy is to analyze its basic structure. In the second, the philosopher investigates the subjective conditions of the possibility of objective reality ... [we ask where are we coming from in seeing it that way]. In the third, subjectivity is re-inscribed into reality ... [our ideological assumptions as to where we are coming from become part of reality].

He provides the example of art: "Reality is not just 'out there', reflected or imitated by art, it is something constructed, something contingent, historically conditioned" (op cit., p. 254). In postmodern art for example, "the transgressive excess loses its shock value and is fully integrated into the established art market" (op cit., p. 256). Similarly, mathematics describes realities that are consequential to past human endeavours or conceptualisations or commodifications.

Mathematics as a field is built in the human's own self-image through its expansion according to a social agenda. Humans, however, are a function of the worlds that they have produced. The mathematical concepts that they have constructed are then built *into* the human self-image. These self-producing and self-validating relationships trap us into thinking that there are universal realities as to what it is to be mathematical. As seen in the last section, mathematics can provide a structuring or formalisation of one's connections to the world. Commodified versions of mathematics have created the illusion that there is something more tangible in mathematical thought that assumes the quality of reality, supporting thoughts directed towards particular arrangements of the world. In Badiou's terms, by counting the elements of a commodity as one, the commodity is brought into existence in the world of school mathematics. This Badiouian approach is discussed in relation to mathematics education by Brown, Heywood, Solomon and Zagorianakos (2012). These constructions become the currency used to measure and classify mathematical thinking. The need for accountability in mathematical learning results in specific transformations of mathematical teaching and learning around commodified concepts.

How then do these concepts provoke our willingness to be governed by them? According to Žižek (1989, p. 43), Althusser "never succeeded in thinking out the link between *ideological state apparatus* and ideological interpellation":

> Althusser speaks only of the process of ideological interpellation through which the symbolic machine of ideology is 'internalised' into the ideological experience of Meaning and Truth. (Žižek, 1989, pp. 43–44)

In our case, the link would be between the assessment structures that govern our practice and our belief in those structures. Whilst we may criticise the structures in theory, our practice is largely compliant. As indicated, mathematics in universities and in schools interpellate individuals, but why? Althusser offers no explanation. Žižek contrasts Althusser with Lacan, who posits some subjective space that exceeds ideological interpellation. In a Lacanian framework, the subjective experience of mathematics can exceed these ideological parameters as a result of individuals *practically* participating in the rituals of schooling. In subjecting oneself to the ritual of institutionalised mathematics one is inadvertently materialising one's belief in it and this belief creates a successful link between *ideological state apparatus* and interpellation. Meanwhile, mathematical thought will always exceed its specific commodified manifestations such as the concepts that are constructed for school and elsewhere. After Kripke, Žižek posits a notion of a "'rigid designator' – of a pure signifier that designates, and at the same time constitutes the identity of a given object beyond the variable cluster of its descriptive properties" (Žižek, 1989, pp. 43–44). The name "mathematics" locates something that is more than the sum of its descriptions, thwarting any consistent account of what mathematics "is". Rather, mathematics is only accessed indirectly through descriptions of the activities taking place in its name. And the sum of those activities is not the whole.

CONCLUDING COMMENTS

The production of mathematical concepts may be helpfully understood mathematics coming into being, or participating in the becoming of mathematics, making it come into being. The learner may experience mathematics as part of herself, a self that is also evolving in the process. Mathematical concepts and the ways in which we relate to them would never finally settle in relation to each other. Their final form always stays out of reach. The building of mathematics then reflects the image we have of ourselves and becomes part of those selves that it reflects. Yet, we may not experience our immersion in mathematical changes in this way. We understand ourselves as operating in a rather more restrictive space decided upon by legislation, by teachers or by expectations beyond our active control. If the world is built in our own image, our children may encounter that world as an external demand out of line with their own perceived needs. Following Hegel, Malabou (2011, p. 24) suggests that the individual "does not recognise itself in the community that it is nevertheless supposed to have wanted …. The individual is 'alienated from itself'". The "self is already implicated in a social temporality that exceeds its own capacities for narration" (op cit., p. 28). These fractures in our self-image can result in adjustments to our tangible reality and to how we encounter it. Mathematics is a function of how we organise its supposed content (concepts, patterns, formulae, procedures) at any point in time. Yet it is also a function of the narratives that report on how we experience it through time, and of the hermeneutic working through of those narratives that generate new dimensions of mathematics (Doxiadis & Mazur, 2012). These narratives may be productive, misguided, manipulative, or functions of particular administrative or ideological perspectives (Lundin, 2012). For example, ideal accounts of mathematics can readily become policing structures in the service of compliant behaviour transforming how subsequent students experience mathematics. Curriculum innovation and associated testing can activate new, perhaps unexpected, modes of mathematical engagement or educative encounters across a community. People or communities more or less identify with these new conceptions of mathematics and shape their practices accordingly. School mathematics does not generally reach for the stars, and often prefers to make do with some rather rusty scaffolding in the name of corrosive metrics. There is a recurrent sense that there should have been more to it than has been allowed. Whilst the Truth of mathematics can sometimes be used to underwrite its ideologically motivated manifestations, we need to trouble the "truths" that are presented to us, towards encountering the spaces beyond and the hold they have on us.

REFERENCES

Badiou, A. (2009). *Logics of worlds*. London: Continuum.
Barad, K. (2007). *Meeting the universe halfway: Quantum physics and the entanglement of matter and meaning*. Durham, NC: Duke University Press.
Brown, T. (2011). *Mathematics education and subjectivity: Cultures and cultural renewal*. Dordrecht: Springer.
Brown, T. & McNamara, O. (2011). *Becoming a mathematics teacher: Identity and identifications*. Dordrecht: Springer.
Brown, T., Heywood, D., Solomon, Y. & Zagorianakos, A. (2012). Experiencing the space we share. *ZDM: The International Journal of Mathematics Education*, 45(4), 561–572.
Deleuze, G. & Guattari, F. (1996). *What is philosophy?* London: Verso.
Doxiadis, A. & Mazur, B. (Eds.) (2012). *Circles disturbed: The interplay of mathematics and narrative*. Princeton, NJ: Princeton University Press.
Feltham, O. (2008). *Alain Badiou: Live theory*. London: Continuum.
Hawking S. & Mlodinow, L. (2010). *The grand design*. New York: Bantam.
Lundin, S. (2012). Hating school, loving mathematics: The ideological function of critique and reform in mathematics education. *Educational Studies in Mathematics*, 80(1–2), 73–85.
Malabou, M. (2011). Is confession the accomplishment of recognition? Rousseau and the unthought of religion. In S. Žižek, C. Crockett, & C. Davis (Eds.), *Hegel and the infinite: Religion, politics and dialectic* (pp. 19–30). New York, Columbia University Press.
Pais, A. & Valero, P. (2012). Researching research: Mathematics education in the Political. *Educational Studies in Mathematics*, 80(1–2), 9–24.
Palmer, A. (2011). "How many sums can I do"? Performative strategies and diffractive thinking as methodological tools for rethinking mathematical subjectivity. *Reconceptualizing Educational Research Methodology*, 1(1), 3–18.
Paton, G. (2012). Numeracy campaign: 17m adults 'struggle with primary school maths', *The Telegraph*, 2 March.
Piaget, J. (1952). *The child's conception of number*. London: Routledge.
Skovsmose, O. (1994). *Towards a philosophy of critical mathematics education*. Dordrecht: Kluwer.
Žižek, S. (1989). *The sublime object of ideology*. London: Verso.
(1997). *The plague of fantasies*. London: Verso.
(2001). *Did somebody say totalitarianism?* London: Verso.
(2006). *How to read Lacan*. London: Granta.
(2011). The spectre of ideology. In I. Szeman & T. Kaposy (Eds.), *Cultural theory: An anthology* pp. 228–244. Chichester: Wiley-Blackwell.
(2012). *Less than nothing: Hegel and the shadow of dialectical materialism*. London: Verso.
Žižek S. & Millbank, J. (2009). *The monstrosity of Christ: Paradox or dialectic?* C. Davis (Ed.). Cambridge, MA: The MIT Press.

PART VI

12

A Relational View of Mathematical Concepts

ALF COLES

Many concepts used in mathematics appear self-evidently to be relations. Speed is a relation between distance and time (which are themselves relations); fractions and ratios can both, in different ways, be seen as relations between numbers or lengths or quantities; volume is a relation between one shape and a unit (typically a unit cube); the gradient of a graph is a relation between 'rise' and 'run'. In this chapter, I aim to show that there is considerable pedagogical advantage in viewing *every* mathematical concept as a relation. I aim to show that by introducing mathematical concepts as relations, the supposed problems for children in abstraction fall away and learning can become fast, imaginative and engaging. The arguments are primarily pragmatic and empirical ones, based on classroom experiences and experiments. The view of mathematical concepts as relations entails a disruption of the notion that learning proceeds through abstraction from manipulating concrete objects to visual models, culminating in abstract entities, and instead suggests a more circular conception of learning and development.

It is not easy to pin down what kind of a thing a relation *is*; I take examples of relations to be differences, similarities and comparisons and wish to distinguish relations from the objects that are being compared or related. I suggest this difference is one of perspective and sometimes choice – it is perhaps possible to view any object (whether real or imagined) as a relation and any relation as an object. An exercise used at the start of art college, when faced with a still life, is to move focus away from drawing the things on the table, to the spaces in between. Instead of focusing on the black text of this page, I invite you to shift attention to the meandering lines of whiteness than run down the page. These are shifts in perception from noticing objects to noticing relations. There is a choice to be made. At moments of insight, we re-see, re-think, re-cognize what we have been attending to. The poet and philosopher Jan Zwicky (2014) suggests, 'all genuine understanding is a form of seeing-as' (§3). Similarly, Gregory Bateson (1979) described the core of his scientific method as the search for a double description of phenomena, where '[*relationships are*] *always a product of double description*' (p. 132; emphasis in the original). When I claim, therefore, that mathematical concepts are relations, this is not an ontological statement, but rather a suggestion that seeing mathematical concepts as relations has advantages, particularly when this is accompanied by a symbolizing of relations rather than objects. I draw on writers who distinguish between abstract and more concrete or empirical modes of thought. What I take this distinction to mean is that abstract thought deals with relations and concrete or empirical modes of thought deal with objects.

The structure of this chapter follows the logic of Walter McCulloch's (1960) challenge: 'What is a number, that a man [*sic*] may know it, and a man, that he may know a number?' (p. 7). To begin, and in order to approach the idea that every mathematical concept is a relation, I take the case of the early learning of number; this leads to a consideration of human cognition more broadly and then to suggestions for how other areas of the mathematics curriculum can be approached in a relational manner.

WHAT IS A NUMBER, THAT A WOMAN MAY KNOW IT?

Mathematician Bass (2015) suggested there are two competing narratives around the development of number sense and its culmination in an awareness of real numbers and the number line. The predominant narrative in schooling is a 'counting world'. Number is introduced as a label for distinct objects, and so, when mapped to an imagined number line, numbers are individual marks. From this point, the number line must be progressively 'filled in', with negative numbers, rational numbers and ultimately, real numbers. Such an approach leads to significant difficulties for students in the jump to the existence of the rationals and then to the existence of the reals. If the basic conception of number is to stand for objects, a disruption to this idea is needed to make sense of fractions. Perhaps another way of saying this is that starting with the notion of number as standing for discrete objects, we introduce number to students in an 'empirical' mode of thought (Goutard, 1964), and this way of thinking is not adequate to conceptualise further developments in mathematics.

Bass' alternative approach to number (and the one I advocate in this chapter) is a 'measurement world', or I would prefer to call it a 'relational world'. Bass draws on Davydov's (1990) curriculum as an exemplar of what it might mean to introduce number not as standing for objects, but as measure. In terms of a number line, whole numbers can be seen as scalings of a unit length. The number line is 'full' from the very start, in the sense that, intuitively speaking, a scaling can get you anywhere on a line. There is no conceptual stumbling block in moving from integer to fractional scalings. Ma (2015) proposed that the basis for arithmetic is an awareness of quantitative relations. In the counting world, this is undoubtedly the case. However, in the measurement world (and for Davydov), before associating specific numbers with specific scalings, awareness can be developed of the broader relations of 'greater than' and 'less than' as the basis for number sense. As well as Davydov, Gattegno (1974) also developed a programme that introduced number not in the context of groups of objects. As far as I am aware, their programmes have never been closely compared, yet, as I hope to draw out in this chapter, there are striking similarities which lead to implications for the whole mathematics curriculum.

Caleb Gattegno (1911–1988), born in Egypt, worked across the world developing a mathematics curriculum based in the use of the Cuisenaire rods.[1] Vasily Davydov (1930–1988) was born and worked in Russia. At around the same time as Gattegno he developed a curriculum that was implemented in schools in Russia. Both educators have their advocates today; for example, Gattegno has inspired the Bronx Charter School in New York and Davydov's ideas are behind the 'Measure up' programme (Dougherty, 2008).

In both programmes for mathematics, children's first lessons revolve around experiences with objects of different lengths. Gattegno (1963) suggests children have some time of 'free play' with the Cuisenaire rods. There is a conviction that children will quite naturally begin making comparisons. Initially the comparisons are 'greater than' and 'less than'. An important step (again in both curricula) is the use of aligning lengths for comparison. And, equally important, if one rod or length is longer than another, then a third rod/length can be added to make the lengths the same. Madeline Goutard worked closely with Gattegno in developing the use of Cuisenaire rods and wrote a book in 1964 in which she detailed her approach. An early task she would do with children, once they had become familiar with the rods and begun making comparisons between lengths, was to focus on the length 'to be added'. This length can be associated with a subtraction and

[1] Cuisenaire rods are cuboids with 1 cm^2 cross-sections and ranging in length from 1 cm to 10 cm. Each length is coloured uniquely (e.g., the cube with 1 cm lengths is white, the rod of length 2 cm is red).

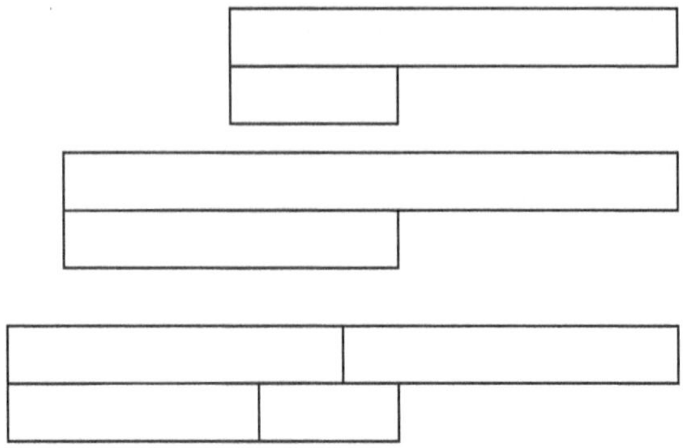

FIGURE 12.1 Families of equivalent subtractions (adapted from Goutard, 1964).

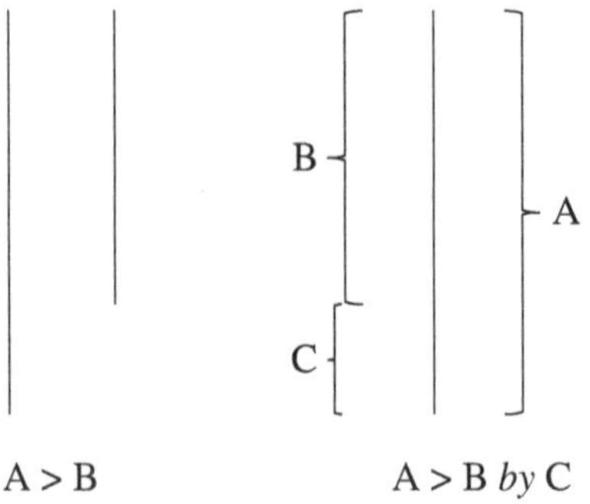

FIGURE 12.2 Comparing lengths (adapted from Schmittau, 2005, p. 19).

Goutard would get children working on 'families of subtractions', i.e., pairs or trains of rods where what is 'to be added' is the same (see Figure 12.1).

Davydov worked with a variety of materials (i.e., there is no equivalent of the central place for the Cuisenaire rods), but the focus, as with Gattegno, was on comparison of measures (whether this be a measure of length, area or volume). At some point, there would also be a focus on 'how much' difference there is between two lengths (see Figure 12.2).

Both programmes would get students in the first grade codifying relations, such as in Figures 12.1 and 12.2, by using letters. Gattegno would label each of the Cuisenaire rods with a letter indicating the colour name, while Davydov would use A, B, C to represent the different lengths. The kinds of statements that students would produce are, for Goutard's families of equivalent subtractions:

$$p = b - y = d - r = \cdots$$

and, for Schmittau:

$$A = B + C$$

$$B = A - C$$

$$C = A - B$$

It is important to note that no numbers are used (on either programme) to represent lengths. In both Davydov's and Gattegno's curricula, number is first introduced as a *comparison* of measurements, when we have the special case that you can use copies of the same length to make a longer length.

For Gattegno, the kind of relationship in Figure 12.3 is used to introduce the number '2' and immediately (as the second number to appear) the number '½'.

$p = 2r$ and the equivalent, $r = \frac{1}{2}p$

(Gattegno, 1963, p. 29).

For Davydov, the role of the 'unit' in measurement is central. From a comparison of lengths, as with Gattegno, the first numbers introduced are in some sense a scaling. Children are encouraged to create their own notation, e.g. tallies, to signify the number of times a unit measure fits into a longer length (see Figure 12.4 for an example).

As Schmittau explained, 'From this, the idea of the counting sequence is developed, as a tool for labelling how many units make up any particular quantity. Then the concept of number is developed as a relationship of quantities.' (ibid). This relationship is expressed as follows:

$$B = 5U \text{ and the equivalent, } \frac{B}{U} = 5$$

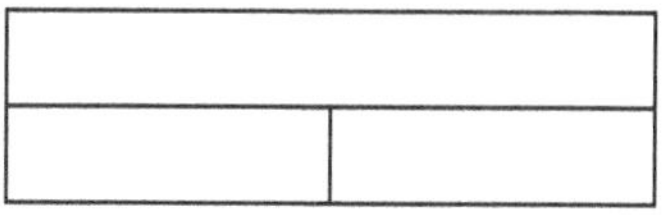

FIGURE 12.3 The introduction of numbers.

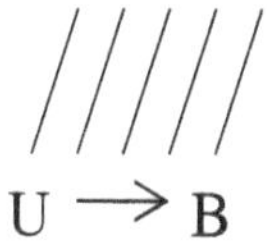

FIGURE 12.4 A notation of comparison (adapted from Schmittau and Morris, 2004, p. 66).

As with Gattegno, this scaling relation is expressed in two different ways, although there is a significant difference in the second expression. Whereas Davydov in some ways 'defines' number as the ratio of lengths ($B/U = 5$), Gattegno stays with the use of number as an 'operator' (or scale factor) and so introduces fractions in the first grade to capture the inverse relationship ($r = 1/2\, p$).

These parallels in the development and introduction of number are striking. What is also clear from both treatments is that number is not being linked to collections of objects, or at least not only to collections of objects. Number is first seen as a dynamic relation. Through introducing number as a relation between quantities, whole numbers become, in effect, scale factors. 5U or 2r mean 'take five of the unit length' and 'take two of the red rod'. Numbers are brought into existence through the action of placing objects against each other.

Gattegno makes use of this relational view of number to work with fractions. For anyone skeptical about whether it really is possible to get children aged six or seven working with fractions, it is only necessary to watch a video clip of Gattegno teaching (http://www.calebgattegno.org/mathematics-at-your-fingertips.html), where in one hour children (who have gained familiarity earlier in the year with the rods) move from being introduced to fraction notation to being able to solve the following questions:

$$\frac{1}{2} \times (36 - 18) =$$

$$\frac{1}{3} \times (18 + 9) =$$

$$\frac{1}{4} \times (9 + 27) =$$

$$\frac{1}{2} \times \left(9 + \frac{1}{3} \times 27\right) - \frac{1}{4} \times 36 =$$

These questions would not be out of place on test items for fourteen- or even sixteen-year-olds in the United Kingdom and yet were tackled confidently by six- and seven-year-olds. Similarly, Davydov developed schematics for working with part-whole relations, and students quite quickly are able to solve problems involving proportional reasoning that would usually only be tackled much later on in a typical UK or US mathematics curriculum.

Students can find the shift into proportional reasoning a significant barrier in learning mathematics. The remarkable idea of both Gattegno and Davydov is that this problem can be avoided by introducing number, from the very start, as a proportional relation, within a 'measuring' or relational world rather than a 'counting' world. And, because the symbols are relations, it seems, furthermore, that the step to considering symbols in relation to one another, leaving behind the objects from which they initially arose, also occurs swiftly for students. It is evident from the preceding discussion that when symbolising (relationally: see Figures 12.3 or 12.4), there are several descriptions possible – viewing the relation from different perspectives. So, from the very start, students are considering relations *between* symbolic descriptions (e.g., the connection between 2 and ½).

An obvious question begged by the relational approach to number and its apparent power is whether the whole curriculum can be treated in similar manner. In thinking about this question, I have found it useful to reflect on what, in these approaches to number, is being symbolized. Clearly, from what has been discussed earlier, I suggest that numerals denote relations (between lengths). However, what is also highly significant is that number symbols are initially used to stand for an action *performed* on the lengths or rods. The relation captured by '2' or '½' in Figure 12.3 is performed by students, in the sense that they place the rods in that special configuration and, from their previous work with rods, likely with the awareness that other comparisons of lengths leave something left over. Also important is that different sets of rods are arranged in the same configuration (2 whites and 1 red; 2 yellows and 1 orange, etc.) so that '2' is not associated with particular rods but with what is common to a range of instantiations of putting two of the same rod precisely against a third. What is *not* being symbolized with a numeral, at least initially, is any actual length.

Reflecting on the effect of these pedagogical moves, it becomes apparent, as I hope to show in the next section, that what happens to learners is not captured by the orthodoxy to a developmental sequence of, say, moving from concrete to visual to abstract. Making this point takes us into questions of what it is to be a human (learning about number).

WHAT IS A WOMAN, THAT SHE MAY KNOW A NUMBER?

The orthodoxy in the United Kingdom in the early years of education and schooling is that children need to work with concrete materials or manipulables to develop their understanding of number. Goutard (1964, p. 3) pointed to this phenomena:

> It is generally agreed that concrete experience must be the foundation of mathematics learning. When children find it difficult to understand arithmetic it is at once suggested that this is because it is too abstract; for small children the study is then simply reduced to the counting of objects.

The importance of the concrete in the learning of mathematics is given a more recent interpretation through, for example, the work of Lakoff and Núñez (2000), who suggested a bodily basis for all mathematical meaning. From this 'embodied' stance, our understanding of number begins with bodily experiences (for example, rhythmic clapping, walking up steps) and we progressively abstract from these experiences, reorganising our perception and culminating in symbolic representations of our actions.

However, there is a well-known paradox linked to this familiar story of the movement from the concrete (perhaps via the 'visual') to the abstract. Piaget (1976, p. 353) summed up the paradox as follows: 'adaptation to the concrete experimental facts is dependent upon the abstract character of the theoretical framework, which allows analysis and apprehension of these facts'. In order to make sense of our perceptions, we need to have some kind of abstract framework within which they fit. Yet we can seemingly only develop these frameworks via abstraction from experience of our perceptions. There is a circularity here: we need abstract structures to make sense of perception and we need perception to build abstract structures (a version of this paradox is the focus of the chapter in this book by Wolff-Michael Roth). Some evidence for the existence of such a circular 'trap' comes from Piaget's experiments with young children. He observed:

> It is obviously not sufficient to make a correct observation of something for it to be accepted, if there appear to be valid reasons for refusing (or repressing) it … if the subject feels that what he [*sic*] sees happen should not have happened, then this observation is not retained or conceptualized. (Piaget, 1976, p. 214)

Piaget observed children ignore results of experiments they had performed accurately (for example, using a simple catapult to throw an eraser into a bin) when it came to describing or predicting what to do. A child might have a 'know-how' and yet seemingly lack the abstract structures needed to make sense of this know-how. Piaget embraced a 'circular relationship between subject and object' and arrived at the view that:

> The subject only learns to know himself when acting on the object, and the latter can become known only as a result of progress of the actions carried out on it ... this explains the harmony between thought and reality, since action springs from the laws of an organism that is simultaneously one physical object among many and the source of the acting, then thinking, subject. (Piaget, 1976, p. 343)

The circularity and entanglement in the emergence of subject and object is a key insight behind the worldview of enactivism (Varela, Thompson & Rosch, 1991), which offers a way through the paradox of how learning and abstraction take place. One difficulty in proposing a viewpoint such as enactivism is that our established patterns of thinking are firmly embedded in the conviction of the separability of subject and object and, as Piaget notes, humans are very good at refusing or repressing evidence that does not fit an existing way of thinking. One way to catch ourselves out, and allow the possibility of a different perspective, is a consideration of illusions. (For Maturana, a key enactivist thinker, the impossibility of knowing, in the moment, whether a perception is an illusion or not led to an insight about the informational closure of living systems; Maturana & Poerksen, 2004.)

Consider the illusion in Figure 12.5. Focus on the two lines of grey tiles; do you see the greys as the same shade?

It seems that humans experience these greys as different. Now turn over the page and look at Figure 12.6, where the same tiles are reproduced, but this time without one column of black. In this configuration there seems to be no difficulty in seeing the grey shading as the same (or at least much closer in tone than before), yet it is the identical grey to that used throughout Figure 12.5. How is it possible to explain the phenomena you have just experienced?

What seems clear from this experiment is that in Figure 12.5, we do not perceive the colours 'in themselves'. This is an example of a family of illusions where it seems as though the context leads us 'astray'. The reason for

FIGURE 12.5 The Munker-White illusion.

FIGURE 12.6 The illusion unmade.

the difference in the way the grey looks in Figure 12.5 has something to do with expectations linked to the pattern of light and dark in the other tiles. In other words, it seems that we do not perceive the grey colour per se, or at least the context of the grey alters how we perceive it. This has always been known to artists, no doubt; the quilter Kaffe Fassett, famous for his bright designs, talks about the importance of dull and grey colours in his design in terms of the overall colour effect (personal communication).

It might be easy to dismiss this illusion as a gimmick, or to imagine that in this instance we simply need to do some alterations to Figure 12.5 in order to see the colours 'as they really are'. However, the illusion suggests a more radical view. To put this at its starkest, since we always perceive colours in a context, we can *never* see colours 'as they really are' – indeed the entire concept of colour 'as it really is' does not even make sense. Therefore, it is not actually the case that we perceive 'colours' as such, rather, we perceive *relations* between colours. Even when we know the illusion it seems impossible to 'see' the greys in Figure 12.5 as the same.

But if we can only make sense of our colour perception of objects through a consideration of context, then it cannot be the case that we really perceive objects at all (what is true of colour will be true of all modalities of perception). Although I may experience the world as a set of discrete objects, if context plays such an important role, then the separation of objects from each other (and hence from me) is really a fiction. Again, this is something that has been well known to artists, and connects back to Piaget's insight about the importance of abstract frameworks in perception:

> [T]o see an object is always to perform an abstraction because seeing consists in the grasping of structural features rather than in the indiscriminate recording of detail. (Arnheim, 1969, p. 68)

We can of course notice details, but when we do, this is preceded by an awareness of more general, abstract and structural features; 'generalities precede particulars in sensory experience' (Arnheim, 1969, p. 221), precisely what Piaget concluded from his experiments. The enactivist conviction (drawing on, for example, Bateson, 1972; Maturana and Varela, 1987) is that the core cognitive function is to make distinctions, and that even the most basic of organisms operate in the world through acting on relations and differences rather than on a representation or awareness of objects, as the colour illusion (Figure 12.5) illustrates for humans. In other words, there need be no paradox in abstraction coming before particulars, when abstraction is taken to mean attention to relations, if our basic mental function is to attend to relations and differences.

RE-CASTING THE CONCRETE—ABSTRACT DIVIDE

If humans perceive relations and distinctions, then the typical developmental sequence of a move from sensori-motor operations to the mapping onto those sensori-motor experiences of more formal operations is put into question (something also critiqued in de Freitas and Sinclair's [Chapter 4 in this volume, Proposition 2]). The relation 'double' (e.g., in Figure 12.3) does not exist *in* either the smaller or the larger rods, but arises through a human comparison between them. In this sense, relations and differences are *always already* abstracted from the objects that give rise to them. If making distinctions is the basic mental function, then whatever problems children have with mathematics cannot be due to difficulties of abstraction. Rather, it may be that the approach to (for example) early number development, in emphasizing objects, is establishing a pattern of thinking about mathematics that makes it difficult.

Goutard (1964) distinguished 'empirical' thought (about objects) from more structural awareness (of relationships). While the objects of mathematics become more and more complex and abstract (for example, we study number patterns that are codified as functions and then treat functions as objects in order to consider their properties, and so on), it does not follow that structural (or relational) thinking is hard, only that the structures about which mathematicians are concerned become nested, one built on another. In the introduction to her book describing her experiences using Cuisenaire rods, Goutard considers this division in ways of thinking:

> It seems to me that there has perhaps been too great a tendency to make things concrete and that perhaps the difficulties children experience spring from the fact that they are kept too much at the concrete level and are forced to use too empirical a mode of thought. (p. 3)

Davydov and Gattegno offer mechanisms for children, from the very beginning, to engage in thinking that is not limited to the concrete or empirical level of thought. It may appear paradoxical that the use of Cuisenaire rods can be talked about, as above, in relation to moving students away from the concrete. And this is where what is being symbolized makes all the difference. If, in using Cuisenaire rods, the white rod is always associated with '1', the red rod with '2' and so on, then the use of the rods will remain in an 'empirical mode of thought' and the entire power of Gattegno's approach is lost. By introducing numbers as relations, abstracted from the concrete context at the very start, there is no concrete-abstract divide to cross, for children. Number symbols (initially '2' and '½') denote a particular configuration of material objects but quickly take on properties in relation to each other. In Coles (2014) I described work with children using a visual image (the Gattegno tens chart) where there was precisely this sense of symbols arising from and for actions and relations within the chart and then children quickly focusing on relations between the symbols themselves.

Some evidence for there being more than one way of thinking about the objects of mathematics comes from recent neuroscientific studies related to early number development. Lyons and Beilock (2013) tested subjects performing basic number tasks such as comparison of size and judgement of whether numerals were in order (4, 5, 6 are in order; 4, 6, 5 are not). What they have discovered is that similar patterns of brain activity are present when comparing the sizes of numerals, or collections of dots and even when judging if three sets of dots are in order of size (perhaps all these are examples of Goutard's 'empirical' thinking), and that a different pattern of brain activity is aroused by being asked to make (ordinal) comparisons of whether three numerals are in sequence (perhaps requiring more structural or relational awareness). Not only this, but the kinds of brain activity used in the latter task are similar to those used in more complex arithmetic and, furthermore, speed at this kind of ordinal awareness is the best predictor (compared with speed at any of the other size comparison tasks) of overall mathematical attainment from grades 2 to 6 (Lyons et al., 2014).

The proposal suggested by the preceding considerations is therefore that the paradox of the concrete-abstract divide in learning number only arises if children are forced to associate numerals too strongly with collections of objects. If, instead, children are introduced to number symbols as relations, then they can be inducted, from the start, into the way of thinking about and with number that is needed for success in mathematics. And what is true of number (that it can be introduced relationally), I aim to show in the next section, is true of every mathematical concept.

ALL MATHEMATICAL CONCEPTS AS RELATIONS

In order to develop an entire curriculum on the basis of mathematical concepts as relations, it is necessary to devise starting points where what can be symbolized are actions and transformations of objects or images. There is not enough space to deal with every aspect of the curriculum, so in this section I take three concepts and show how they can be thought of as relations and how, in doing this, some reordering of the traditional order of curriculum topics may be required. These three concepts are chosen as illustrative and include one example each from primary, secondary and post-sixteen phases of schooling.

Area as a Relation

The concept of area can be considered in an empirical manner, with concrete objects or visual images, and it is a concept that children can have difficulty with. However, 'area' is an inherently relational concept when considered as a relationship to a unit. In a similar manner to the way that Gattegno introduces the symbol '2' to represent 'double' rather than two objects, an area of '2' can be introduced to represent the situation of one shape fitting into another shape twice (and hence also the first shape has an area ½ the second shape). A relational approach to area would not be so much concerned with attaching numerical values to the sizes of shapes as in comparing them – Which are bigger? Which are smaller? – and then considering the case where one shape can fit an exact number of times into another shape. In essence, by considering area as a relation, the concept arises out of the wider mathematical structure of transformation geometry and, in particular, enlargements. It might be, therefore, that we work on enlargements with students before area (a reverse of the typical sequence). It seems, at the least, that there are indeterminate relations between these concepts, and ones we typically see as necessary to build on others are not definitively so; any curriculum needs flexibility, therefore, to be responsive to students' awareness. It would be possible to use awarenesses about area to work on enlargements but equally possible to use awarenesses about enlargement to work on area.

Algebra as Relation

In the book *Starting Points* (Banwell, Saunders & Tahta, 1986), there is the proposal that whenever functional relationships are discovered by children

TABLE 12.1 *A function representation*

3	6
5	10
6	12

or being considered in a lesson, there is always the same way of representing them, in a tabular form (Table 12.1).

One way of setting up this representation is through the 'function game' (also described in Brown & Waddingham, 1982). The teacher has a rule in mind ('doubling' in the case of Table 12.1) and writes out two 'inputs' with arrows to the 'output', and then writes a third input and invites the class to guess the output. This game can be powerful when played in silence, with the teachers (and later perhaps a student) indicating if the output is correct with a ☺ or ☹. At some point, when most of the class have figured out the rule, the teacher might put 'n' as the input, to invite a sharing of these rules. Even with a rule as simple as 'doubling' there may be differences in how students were applying the rule, that can be captured in their algebraic expressions (e.g., $n+n$, or $n\times 2$, or $2n$). The teacher can also put 'n' in the 'output' column, with the arrow backwards, and invite expressions for the inverse of the rule they have just found.

Algebra, in this treatment, represents a relationship between two sets of numbers, and working on functions can provide motivation to consider more routine or technical algebraic techniques (for example, in showing why all the different algebraic rules are the same). In a standard UK curriculum, rather than an introduction to algebra, functions would typically appear much later.

Complex Numbers as Relations

A complex number can be seen as a relation if it is considered as a function that transforms shapes. Because complex numbers need two dimensions to represent them, to get a sense of how they might transform a shape, we need two sets of (2D) axes. Figure 12.7 shows how a circle is transformed under the operation $z \to 2z$.

The prerequisite to considering complex numbers as relations is to know i as the square root of negative 1, and to know how to express complex numbers on an Argand diagram. Starting with the left-hand diagram, students can be invited to conjecture what the shape would be transformed to if both real and imaginary parts were doubled. For example, i, at (0, 1) on the diagram above, will be mapped to $2i$, at (0, 2) on the right-hand diagram.

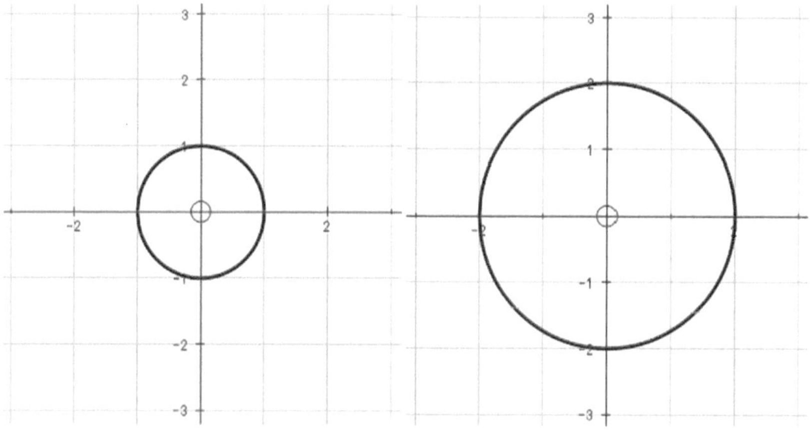

FIGURE 12.7 A representation of $z \to 2z$.

From my own experience of teaching, students can quite quickly become independent in testing other complex transformations and generating their own pairs of images. In each case, students can be invited to predict and test what transformation will be described. A potential mathematical appreciation on offer in this task is around the relations between one entire system (transformation geometry, e.g., translations, rotations, reflections, enlargement) into another system (complex numbers). It is also possible to work on the relations between the transformation entailed in a matrix $\begin{bmatrix} a & b \\ c & d \end{bmatrix}$ and an equivalent complex number (e.g., $\begin{bmatrix} 0 & -1 \\ 1 & 0 \end{bmatrix}$ is a rotation of −90 degrees around the origin, and is therefore equivalent to i). Again, the sense of complex numbers as transformations would typically appear late in any course that introduced complex numbers. The task above can set up the ideas of the geometrical interpretation of multiplication by complex numbers, leading to de Moivre's theorem, viewed as a statement about transformations.

DISCUSSION

The three examples presented and the case of number discussed earlier indicate how a selection of mathematical concepts can be interpreted as relations. It is the proposal of this chapter that *every* mathematical concept can be treated in a similar manner. As has been clear in the four examples, the view of mathematical concepts as relations disrupts the typical sequence of the treatment of ideas, generally by familiarizing students with a wider 'whole', out of which particular concepts arise. To generate starting points that allow a relational view of mathematical concepts, there is a need to focus on broader (mathematical) structures. There is an important role for visible or tangible objects (Gattegno, 1974) but also a danger if these are used to symbolise concepts too directly. If a concept is to be approached relationally, then images or materials are needed where what can be symbolised are actions on or relations between the materials, or within the images.

As a mathematician, if I am presented with an expression such as $2x^2 + 7x - 9$, a whole host of associations arise, I may recognise the statement as an expression that can be factorised, I may picture the graph, or the quadratic formula may come to mind. Depending on what I may be asked, or decide, to do next, particular associations will be foregrounded and others will fade. The concepts and associations I invoke primarily gain their use and meaning from their links to other concepts. If concepts have been introduced as relations, then a symbol's links to physical objects or images take their places as just some among a myriad of connections. When introduced as relations, children can access their imaginations when working with symbols, as mathematicians do. It is the perspective of this chapter that difficulties in mathematics are much more likely to arise from children not having access to any imaginative response to symbols (such as my reaction to the equation above) rather than any supposed lack of capacity for abstract thinking (thinking that is amply demonstrated by anyone who has taught themselves their first language). Symbols are meaningful to mathematicians but not in a direct ('this' means 'that') manner. It is in part the ambiguity and flexibility of mathematical symbols that give them power and use; this is the game of mathematics, but it is a game that many students are not let in on.

Gattegno and Davydov achieved extraordinary results when using their curricula. The pedagogical advantages of approaching concepts as relations, as has only been touched on in the examples above, include the following:

- inverse operations can be considered simultaneously (e.g., fractions with integers; fractional enlargements with standard enlargements);
- there is no disruption as ideas become more complex; relational images or representations will not need to be 'unlearnt', for example in the move from integers to fractions;
- there is something the symbols 'mean' and can point to, as well as being abstract from the beginning;
- linked to the point above, symbols can trigger the mathematical *imagination*, i.e., there is something that can be evoked by mathematical symbols if they are introduced as relations;

- symbols can quickly take on relations to each other; because they are introduced in a complex whole, there are other symbols they *can* relate to;
- complexity can also be limited in the beginning to allow a gradual development, while still being 'abstract' from the start.

Relations are at once material (arising from a consideration of objects) and abstract (the relation of, say, 'double' is not linked directly to any object). They exist neither in the objects themselves nor in any human mind, but rather arise from the interaction of humans with the world and each other. Concepts are never fixed: each use adds a different complexion to the web of connections that surround it and changes us as humans, in the kind of circular relationship suggested by Piaget.

REFERENCES

Arnheim, R. (1969). *Visual thinking*. Berkeley, CA: University of California Press.

Banwell, C., Saunders, K., & Tahta, D. (1986). *Starting points: For teaching mathematics in middle and secondary schools*. St Albans, UK: Tarquin Publishers.

Bass, H. (2015) Quantities, numbers, number names, and the real number line. In X. Sun, B. Kaur, & J. Novotna (Eds.), *Proceedings of the twenty-third ICMI Study: Primary Mathematics Study on Whole Numbers*, (pp. 10–20). Macau, China: University of Macau.

Bateson, G. (1972). *Steps to an ecology of mind*. Chicago: University of Chicago Press.

(1979). *Mind and nature: A necessary unity*. New York: E. P. Dutton.

Brown L., & Waddingham, J. (1982). *An addendum to Cockroft*. Avon, UK: RLDU.

Coles, A. (2014). Transitional devices. *For the Learning of Mathematics*, 34(2), 24–30.

Davydov, V. (1990). *Types of generalization in instruction: Logical and psychological problems in the structuring of school curricula*. Reston, VA: National Council of Teachers of Mathematics.

Dougherty, B. (2008). Measure up: A quantitative view of early algebra. In J. J. Kaput, D. W. Carraher, & M. L. Blanton (Eds.), *Algebra in the early grades*, (pp. 389–412). Mahwah, NJ: Erlbaum.

Gattegno, C. (1963) *Mathematics with numbers in colour Book 1: Qualitative arithmetic, The study of numbers from 1 to 20*. Reading, PA: Educational Explorers Ltd.

(1974). *The common sense of teaching mathematics*. New York: Educational Solutions Worldwide Inc.

Goutard, M. (1964). *Mathematics and children: A reappraisal of our attitude*. Reading, PA: Educational Explorers Ltd.

Lakoff, G., & Núñez, R. (2000). *Where mathematics comes from: How the embodied mind brings mathematics into being*. New York: Basic Books

Lyons I., & Beilock S (2013). Ordinality and the nature of symbolic numbers. *Journal of Neuroscience*, 33(43), 17052–17061.

Lyons I., Price G., Vaessen A., Blomert L., & Ansari D. (2014). Numerical predictors of arithmetic success in grades 1–6. *Developmental Science*, 17(5), 714–726.

Ma, L. (2015) The theoretical core of whole number arithmetic. In X. Sun, B. Kaur, & J. Novotna (Eds.), *Proceedings of the twenty-third ICMI study: Primary mathematics study on whole numbers*, (pp. 34–38). Macau: University of Macau.

Maturana, H., & Poerksen, B. (2004). *From being to doing: The origins of the biology of cognition*. Heidelberg, Germany: Carl Auer.

Maturana, H., & Varela, F. (1987). *The tree of knowledge: The biological roots of human understanding*. Boston, MA: Shambala.

McCulloch, W. (1960). Alfred Korzybski Memorial Lecture. *General Semantics Bulletin*, 26/27, 7–18.

Piaget, J. (1976). *The grasp of consciousness: Action and concept in the young child*. London: Routledge & Kegan Paul.

Schmittau, J. (2005). The development of algebraic thinking: A Vygotskian perspective. *ZDM, The International Journal of Mathematics Education*, 37(1), 16–22.

Schmittau, J., & Morris, A. (2004). The development of algebra in the elementary mathematics curriculum of V. V. Davydov. *The Mathematics Educator*, 8(1), 60–87.

Varela, F., Thompson, E., & Rosch, E. (1991). *The embodied mind: Cognitive science and human experience*. Cambridge, MA: The MIT Press.

Zwicky, J. (2014). *Wisdom & metaphor*. Edmonton & Calgary, AB: Brush Educational.

13

Cultural Concepts Concretely

WOLFF-MICHAEL ROTH

> How do children learn to see 'another bird' when they saw a bird once, or to re-see a car or a friend? It's a very obscure question, how it is that persons learn to see generalities, or see objects again, or see 'another' of an object. (Sacks, 1992, p. 30)

Historically, concepts have been defined as mental entities that allow thinking to occur. A concept would then be a representation of something based on a commonality of a class of objects. In the classical, constructivist approach, empirical concepts are derived from experience by means of abstraction (Kant, 1956). *But how would an individual who does not already know what is cultural about objects encountered sequentially come to abstract precisely those features that make some of the objects members of a cultural concept while excluding others?* This is the central question of this chapter. Sacks (1992), in his work, does not take concepts ("generalities") to be constructed in the solipsist mind, but assumes social processes in which concept words have important functions. This is consistent with a position that holds that "the relation between higher psychological functions at one time was a real relation between people" (Vygotskij, 2005, p. 1021). If concepts exist as real relations with others first, then they are grounded both materially and relationally, a characteristic move of Vygotskij's concrete human psychology.

To work out this take on the nature of knowing generally and on the nature of concepts particularly, I pursue a different route to theorise concepts by drawing on the idea of a documentary method of interpretation (Roth, 2015). Accordingly, we never know cultural objects other than through their concrete manifestations by means of which they document themselves. This approach allows us to theorise concepts as classes of concrete experiences that we refer to by a common name, but for which we may not have, and do not even need, the capacity to state a rule. Thus, for example, we know what a cup is without ever having had to learn a rule of what makes  but not  "a cup" (concept), and how we know that an object is "a cup" rather than "a mug." We never acquire or require a concept of cup in the classical sense, which is based on a property that is shared by all. Instead, the cultural object *cup* is relevant in a myriad of concrete experiences associated with equally concrete objects that are or may be referred to by the same word "cup." *Cup*, then, pertains to an inherently open, changing class of concrete experiences. This leads us to concepts as (concrete) wholes given through their concrete parts, which are parts only because of the wholes that they constitute. Concepts thereby are never abstract entities, but always and inherently concrete. One consequence is that there cannot ever be a difference between knowing a concept and applying a concept; another consequence is that concept learning requires more than exposure to a few instances. Because many philosophers have used the cube as an example of how we know (mathematical) objects and concepts, I use examples from a second-grade mathematics classroom where students learn, among others, about what makes «a cube» "a cube."

LEARNING TO "DO [CATEGORIZING ACCORDING TO GEOMETRICAL PROPERTIES]"

In this section, I present empirical materials from a unit on three-dimensional geometry in a second-grade class. I begin by describing observations made later in the unit, which exhibit conceptual understanding, and then turn to the first lesson in the course where we can find the origin of the concepts. The two teachers involved had announced to the children that they would be doing a unit on "geometry." The account also provides an explanation for how we come to act in accordance with (not *caused by*) rules without knowing rules or the domain that the rules govern.

Individuals Do and Account for Classification

In a curriculum designed on enactivist principles, eight days into the unit, I recorded the following episode. In this episode, the students completed a task that required them to shape a blob of plasticine into the shape of a mystery object that was positioned inside a shoebox where it could be felt but could not be seen because of a screen. In one group, two girls (Jane, Sylvia) soon had fashioned rectangular solids, and the third girl, Melissa, had a cubical shape. She was repeatedly asked directly why she thought it was a cube, and also provided an account in the exchanges with her two peers.

Fragment 1

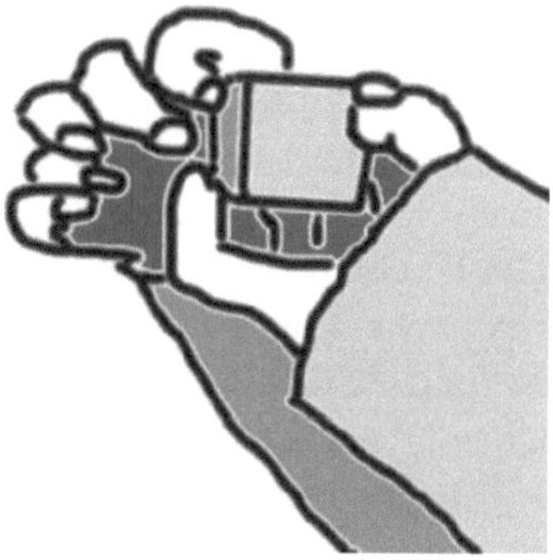

01 M: feel it eh? (1.0) i have felt it's a cu:be (1.5) hU:::::::ge.
02 (1.5) ((*Jane grimaces, questioningly?*))
03 S: no it's not a cube ((shakes head while rH in box))
04 (2.3)
05 J: i didn't feel a cube.
06 S: me either.
07 (4.3)
→ 08 M: I did. (0.6) I I checked the sides like that. ((*Caliper configuration, see right, held to 3 different sides.*))

Melissa states having felt a "cube," whereas Jane and Sylvia state that they have not. Melissa restates that she did (feel the cube) and provides an account for the procedure she used. Forming a caliper configuration with her thumb and index finger, she holds it to each of three edges orthogonal to each other (turn 08). Here, she provides not only a statement but also, as scientists do, a description of the method by means of which the observation was derived. Her peers are not convinced. And a little later, Jane states, "If I feel the top, it seems like its square but if I feel the side it seems like rectangle." She, too, provides verbal accounts for the ways in which two different parts of the mystery object felt: from the top, it feels like a square; from the side, it feels like a rectangle.

In this instance, readers will agree that these children can be said to know the concepts of cube and rectangular prism (cuboid), evident in the way they classify perceived (seen or felt) objects with a defining logical predicate. At the outset of this study, however, that is, for example during the first lesson, the children did not classify objects according to *geometrical* properties, but instead put them together in terms of size or colour. Moreover, out of twenty-two students, twenty-one categorised objects and provided a justificatory account only *after* the teacher had requested: "Tell us your thinking."

In the classical concept learning paradigm, participants are exposed to a number of cases that are marked to be examples of a concept and other cases marked to be counter-examples. Individuals are asked to state the rule that distinguishes examples from non-examples of the concept. But in life, we do not learn concepts in this way (Sacks, 1992). We learn what a queue is by participating in forming queues, one at a time, and each one comes to be what it is based on whatever situated rule is in use. This is in part the reason why exceptions to rules do not lead to a change of concept. How then do children learn concepts typical of mathematical culture when all they have available are the relations with others, and when the very idea of the individual building cultural concepts from the bottom up and in his/her mind leads to contradictions? We do not learn to appropriately participate in queues by constructing a concept of a queue and then testing it for viability in subsequent experiences. That view, where the individual is at the origin of what there is to know, is a myth that numerous scholars within the societal-historical tradition of the social sciences denote by means of the term *Robinsonade*. Rather than speculating about or mobilising worn theoretical terms to account for the learning of concepts, I offer the following observations from the second-grade mathematics course.

The Basic Machinery: Tying Classificatory Action to an Account

In their first lesson, children were asked to sort objects that they were pulling from a black plastic bag. The fundamental problem that the children have to solve is sorting the objects in a way characteristic of geometry without knowing what geometry is or what is particularly geometrical about the objects. They are invited, after the teacher has placed a first object on a coloured mat, to pull an object from the bag and then either place it with an existing group or to create a new group on an empty mat available. The teachers state that the grouping may not be based on colour or size. All twenty-two instances of sorting follow the same basic pattern, which, depending on the particular answers, was expanded because some form of repair was required – e.g., requesting to reconsider placement or reason for the placement – until the object at hand had found a place. This place was part of a(n) (perceptual) order that was only emergent for the children but that would be consistent with the geometrical properties of the objects. This basic pattern, "the machinery," involves tying a categorisation move to a verbal account (reason, thinking). Following Garfinkel and Sacks (1986), we may characterise the work that was thereby accomplished in the form "doing [sorting according to geometrical properties]," where the nature of the adjective "geometrical" was itself the result of the task as a whole. (Teachers also used the verbs "to categorise" and "to group" as synonyms.) In practice, the basic pattern of the work took a form apparent in Fragment 2.

Fragment 2

01 N: ((*Nathan pulls an object from the bag, places his object on the pink piece of paper,* sees *offprint, and starts walking back to his seat on the floor.*))
02 W: now tell us your thinking.
03 N: um this one is like square.
——————————> ((*Nathan picks up the small brown cube and places it on top of the yellow cube, then places it next to it.*))
04 W: Anybody else want to add to that?

In this fragment, we observe, as the offprint shows, Nathan place and thereby categorise an object, followed by Mrs. Winter asking him to tell his thinking. She does more: in asking him, she also makes available the assessment that merely placing the object on a mat is not enough. Nathan accepts the invitation with the reply, "this one is like square." Mrs. Winter then invites others to add; again, she does more, because she in so doing "accepts" the preceding reply. As the remainder of the session shows, had the placement or the account been incorrect, she would have first invited Nathan, whose turn it was, to repair. She did so in twenty-one of the twenty-two cases; and only the twenty-first student immediately began to reason as he compared his object with those in existing groups. Thus, in every situation during this lesson where the reply was not consistent with what adults know to be geometry, some form of action was initiated that allowed students to repair their classificatory action. In each of these repairs, students also were asked to provide a reason – that is, to make the grouping *tellable*. If a student did not arrive at producing such a telling, then others were invited to do so. The next turn at pulling an object was not initiated prior to having arrived at the production of such a verbal account. In this, students were invited to provide a verbal account of the classification. It is precisely this account – or, more generally, the tellable nature of acting in one rather than another way – that distinguishes classification and knowing a concept in mathematics from knowing a concept in many other mundane situations. For example, in a fish hatchery, fish are sorted without requiring a verbal account of why a specimen is to be laid on one rather than another conveyor belt; but if an error is detected, it will be corrected (Roth, 2005).

In both types of instances, we observe work with a structure of this kind: "doing [sorting by X]," where X in the case of the fish hatchery pertains to size and shape of the specimen's head and in the case of the mathematics class pertains to a geometrical property. Although Nathan initially does precisely what the fish sorters in the hatchery are doing, it is not sufficient. It is important that he not only place the object with other objects to make them part of the same group; this order has to be tellable. But tellability fundamentally is social, for whatever Nathan can intelligibly say is already understandable by others as well. Whatever geometry they arrive at as a result of this lesson, it is social not because it has occurred in a group setting – which is a trivial sense of "social" – or because the entire classification of twenty-two mystery objects was achieved by the collaboration of students and teachers. Rather, it is social because the very organization of the lesson and the tellability of the objects is social *prior to anything that one can say to have been constructed by the group or some individual.*

What the students learn, in and through their participation, is the work of "doing [classifying according to shape in a tellable way]." They do so by working through twenty-two mystery objects, and, as a class including the teachers, an ordered system has evolved consisting of classes of objects (including cubes, rectangular solids, pyramids, spheres, and cones). Each class is defined by its membership – that is, each object is part of a group because it meets the class property – and, simultaneously, the class is the result of the properties of each member. As a result, the students come to know each of the classes incrementally. Each new member (re)constitutes the class, as defined by the membership, and each new member is accepted because of some purported (checkable and tellable) property. A concept then is not something abstract. Instead, whatever its name denotes is associated with the manifold of concrete ways in which it manifests itself. Each potential new member may be interrogated for fitting with *this* or *that* existing class, or, alternatively, it may require the formation of a new class.

The structure of this machinery is of the kind that Vygotskij (2005) intimated to be at the heart of societal-historical (re)production. His example was pointing, which begins with a child's failed grasping movement (action), which the mother takes to be pointing by giving the child the object that was lying in the extension of the arm, and finally the child begins to point on its own. In our situation, it is both the invitation to expand the action by tying a verbal (explanatory) account to it and the ratification of the placement/accounting pairing, which accept from the many actions of an individual child precisely those that conform to the cultural conception of the object.

From Social Relation to Individual Concept

How would the epistemological Robinson Crusoe – if he were indeed untutored, which he was not – know what is social in what he perceives? Just as the children are classifying by colour and size, without initially seeing that there are other possibilities for classifying, Robinson could be classifying according to different categories. This is precisely what Vygotskij and Luria showed in the difference in colour categorisation among Uzbek peasants and schooled individuals. However, even classifying by colour and size is already social, which children learn in the same ways and at a much earlier stage in their life than classifying according to geometrical properties (Roth, 2015). In the "epistemological Robinsonade," the individual constructs his/her mathematical knowing from the bottom up, thought of as a psychic function of the individual, an element of *society* (Davydov, 1988). Having realised that it would be nearly impossible for the individual of our times to reconstruct everything cultural from the bottom up, many scholars have turned to social constructivism without nevertheless abandoning the Robinsonade of individual constructivism. In the social version, which in school contexts always is incidental, the construction is said to first occur in the group before it occurs on the part of the individual who internalises whatever there is to internalise. The answer and counter-position to the epistemological Robinsonade can be found in the work of Vygotskij (2005). In the opening paragraph, I quote the scholar's diction that higher psychological functions and the relations between them first are real relations between people. He does not say that the functions first are *in* the social relation, which is a trivial statement given that the children are in school with their teachers. Instead, he writes that the function *is* a relation first. This can actually be observed in the fragment above, as shown in the following considerations.

Initially, the pairing of the action of placing the object on a mat together with a verbal account exists *as* the Mrs. Winter–Nathan relation. We see this clearly when the relevant turn pair is transcribed in a manner that also takes into account the hearing:

01 N: ((*Nathan places object*))
02 W: now tell us your thinking.
03 N: *now tell us your thinking.* um this one is like square.

Seeing that Nathan is walking back to his seat, Mrs. Winter responds to Nathan's action by inviting him to "*now* tell us your thinking." That is, Nathan first has placed his object and *now* is invited to tell his thinking, which he accepts in providing a verbal account. In so doing, the placement comes to be tied to the verbal account. But that tie involves Mrs. Winter as well. The statement "now tell us your thinking" not only belongs to Mrs. Winter, whose vocal cords and mouth resonate to produce the sound, but it also belongs to Nathan, in whose ears the sounds are received, ring, and lead to the recognition of the words. So it is not correct to say that Nathan internalised something that first existed outside of him: he acts, actively listens, talks. There is nothing outside that is not also already inside – which is precisely the late-Vygotskian Spinozist position on learning and development (e.g., Mikhailov, 2001).

The classification (as per placement of the object with another cubical object) and the verbal account are the result of joint (social) work. That work also makes the relation of Mrs. Winter and Nathan. The exchange constitutes the relation, and it is as such that the pairing of placement and account (reason) here exists. What is incidental to the situation is the fact that there are two individuals, and this coincidence does not constitute the concept as social. That exchange relation itself is a manifestation of the social, especially the tellability of certain phenomena, and it is because of the tellability that the concept is a social object in a much more profound way than that there was a group of students and a pair of teachers present and contributing to the ultimately observable arrangement. The tie exists *as* the relation between people first.

One characteristic of the culture of mathematics is that its members act in *recognisable* ways, and they orient to acting in ways that indeed are recognisable on the part of others (Livingston, 1986). Doing observably and recognisably, mathematics requires this pairing of actions and accounts, whereas it is not required in the hatchery, where the action of properly placing a fish specimen is sufficient for the practice. The placement of the specimen itself is an account that categorisation has occurred, and the placement is the appropriate account such that when seen it suffices for another sorter to recognise her as a member of salmon-sorting people. In the context of the mathematics lesson, we observe how an aspect of mathematical culture first exists as a social relation between people before it exists for each individual child, as shown in the first subsection. That relation itself is built on a sociality that precedes the social construction of concepts or socio-mathematical norms: the words "now tell us your thinking," as any other word spoken in the lesson, always already is the result of the social, because a sound is not a word unless it is a possibility for two (Vygotskij, 2005). This is apparent in the turn 02 / turn 03 pair, where the "now tell us your thinking" is a reality for Mrs. Winter and for Nathan, or it is nothing, a failed attempt at communicating. It would be followed by repair actions, such as when Mrs. Winter reformulates what she says repeatedly until the student finally replies.

THE DOCUMENTARY METHOD: CONCEPTS CONCRETELY

In these lessons, the children do not learn to classify – at least not from the bottom up. They already know to classify competently, based on size and colour. What they do not yet do at the beginning of the lesson is classify a collection of object *geometrically*, that is, in ways that reproduce observably geometrical categories. In the first lesson, there are twenty-two cases, each associated with a different turn attributed to a specific student "whose turn it is to" speak, classify, and provide reasons. The turn comes to a close when others have had an opportunity to add. In other lessons, students are invited to collect objects in the classroom that go with an object that they have received or drawn from a bag. Here, some of the objects may be the same as they have encountered during the first lesson, and others may be new objects that thereby come to be members of a class that their membership contributes to constituting. Learning to do another one, to recognise a concept or to provide an ordered collection of objects that represents some concept, includes orienting in ways such that one accounts for *why*, to (be able to) provide a reason. In school, as in many other situations, "for no reason" is treated as not a legitimate answer. In coming to act in ways that are recognisably ways of acting in this classroom and to orient in ways so that their action be recognisable in such ways, the children are taking their first steps to participate in (school) geometry. Earlier I described the basic machinery and work by means of which acting in recognisable ways is coming about: tying (a) an action of placing an object with others or finding a placement on its own with (b) a verbal account of the reasons for such placement.

How would a concept of that kind and learned in this way come to be used? A tentative answer already has been suggested: the *documentary method* (Roth, 2015). The idea of the documentary method originally has been developed in a very different context and for solving a very different problem: how a researcher knows and investigates a cultural phenomenon such as the concept of Weltanschauung. It has subsequently been proposed as the ways in which human beings more generally find out about something that they do not already know. As a means for investigation, the method works like this. The researcher takes some observable fact – including verbal statements – as a concrete manifestation of the phenomenon of interest. A painting, a piece of music, or a novel produced during a particular year all may be taken to be concrete manifestations of the reigning zeitgeist. That zeitgeist is known only through its concrete manifestations. These manifestations do not have (to have) some feature in common that could be abstracted from them as the essence of zeitgeist, in the way that Kantian concepts function. Instead, zeitgeist is known concretely in and through its manifestations. The method is used not only by social or cultural scientists but is a mundane method of inquiry, such as when we take what someone says as a manifestation of their thinking.

Mathematical cognition functions in the same way, for example, in the way we know and perceive cubes (Merleau-Ponty, 1945). We never actually see a cube in its totality, the six sides, eight vertices, or twelve edges; and even if the cube were to be made of glass so that we can see through it, the sides and edges would not appear of equal length or equal sides, and the angles would not appear to be 90 degrees. Instead, we perceive cubical objects in perspectival ways, inherently and without reflection knowing how the object, if it is a cube, will change appearance in a specific manner when we move with respect to it. This movement may be of the object itself, or our movement around the stationary object. The unity of the object, which is known only through its appearances, derives from the unity of the perceptual body. Not only is the unity of the cube not perceivable; it is also *inconceivable*. Instead, "the cube with six equal sides is the limiting idea whereby I express the carnal presence of the cube which is there before my eyes, under my hands, in its perceptual self-evidence" (Merleau-Ponty, 1945, pp. 236–237). As a consequence, the sides are not perceived as projections that we have to interpret to infer the object to be a cube. Instead, what we perceive precisely are the sides of a cube. When Melissa – as others in her classrooms in the later stages of the curriculum – uses the caliper configuration, holding it to three respectively orthogonal edges after making 90 degree turns to describe the procedure she used for determining the nature of the mystery object as a cube, she acts out precisely what Merleau-Ponty notes to be the way in which cubes are known – as sequentially ordered intersubjective perceptions. Here, rotating the cube in her hand in specific ways yields the same perception (measure) of length of the edge. That is, the cube is known and recognized through its concrete manifestations and through the ways in which these manifestations change when the cube moves (in our hands) or when we move with respect to it. Most importantly, and thereby countering any present-day constructivist argument, "I *do not construct* the geometral projection that accounts for these perspectives, but the cube is already there in front of me and reveals itself through them" (Merleau-Ponty, 1945, p. 237, emphasis added). In this description, we precisely observe the cultural object "the cube" and its concrete manifestations (documentary evidence) through which it is known.

FORMALIZING CONCEPTS

In this chapter, I offer an approach to concepts that differs from the standard constructivist account. The difference between the two ways, the Kantian and the documentary method, can be articulated mathematically. In the Kantian (common) case, the concept is defined by some property that is common to all members of a category independent of all the other properties that these members have and that might vary considerably (e.g., the color and size of cubes differ). The concept C then can be stated in terms of a set-builder notation: $C_K = \{x \mid \Phi(x)\}$, where $\Phi(x)$ is some logical predicate that defines the abstracted property. Once the generator is known, the concept is fixed. The documentary method, however, works with concrete manifestations without requiring that there is a common property. In this case, the concept C_{DM} is constituted by the increasing set of concrete instances a, b, … $C_{DM} = \{a, b, \ldots\}$. Learning a "concept" occurs case by case, ordering material instances such that a concept exists only as a class of concrete instances to which new objects can be associated. Each instance is a manifestation of the class, which exists only as a whole from parts, and is inseparable from the ensemble of the parts. Each object will be representative of a class of which it is a member. The work they do is establishing membership in classes, which initially do not exist (for the children), classes that exist only through the membership of other objects, and the representativeness of which may be expanded with each new member.

When a new object is like "this," then it is to be taken like the objects of an existing class. In this designation, it acquires and can be justified in terms of existing accounts. As a number of objects have been designated in this way, a class is constituted and called "cubes or squares." Each object, initially a mysterious ad therefore unclassified one, has found its place because it is like all others that can be designated in the same way. There then comes a point where the new object is classified "at a glance." Before that there might be questions such as "is it consistent with what we said about that group?" At that point then, the classes, "square and cube," "pyramid," or "cylinder," allow seeing mathematically what there is to see in the context of mathematics. Once an object even tentatively is assigned membership with an existing class, it is assigned all the other properties of members of the class without checking. But part of the accounting procedure may in fact involve asking, as Mrs. Winter often did, "Does it fit with what we said about this group?" The procedure takes case by case, tentatively placing it with one (sometimes moving about doing it with every group) and asking, "Does it fit?" or "Same or different?," and then account for fit or misfit in terms of an item that fills the slot in "because ..." In the end, there is a collection of dissimilar things, in colour, size, etc., all associated with the same name; and they become one – at least, once they are treated as one in other sort of situations, the child will be said to have learned the concept.

REFERENCES

Davydov, V. V. (1988). The concept of theoretical generalization and problems of educational psychology. *Studies in Soviet Thought*, 36, 169–202.

Garfinkel, H., & Sacks, H. (1986). On formal structures of practical action. In H. Garfinkel (Ed.), *Ethnomethodological studies of work* (pp. 160–193). London: Routledge & Kegan Paul.

Kant, I. (1956). *Werke [Works]*. Wiesbaden, Germany: Insel.

Livingston, E. (1986). *The ethnomethodological foundations of mathematics*. London: Routledge & Kegan Paul.

Merleau-Ponty, M. (1945). *Phénoménologie de la perception* [Phenomenology of perception]. Paris: Éditions Gallimard.

Mikhailov, F. T. (2001). The "Other Within" for the psychologist. *Journal of Russian and East European Psychology*, 39(1), 6–31.

Roth, W.-M. (2005). Making classifications (at) work: Ordering practices in science. *Social Studies of Science*, 35, 581–621.

(2015). *Concrete human psychology*. New York: Routledge.

Sacks, H. (1992). *Lectures on conversation* (vol. 1). Malden, MA: Blackwell.

Vygotskij, L. S. (2005). *Psixologija razvitija čeloveka [Psychology of human development]*. Moscow: Eksmo.

PART VII

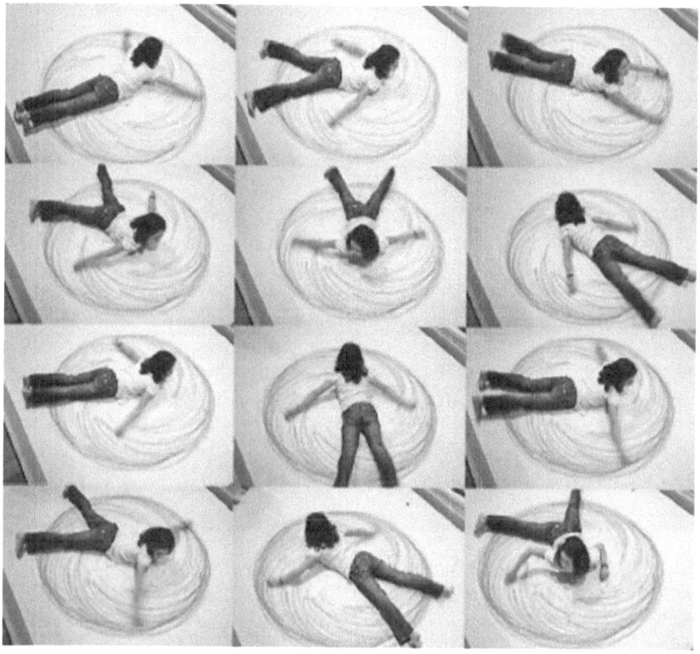

14

Ideas as Species

BRENT DAVIS

A few centuries ago, the title of this chapter would likely have been read as nonsensical – and not because it seems to be offering an unfamiliar metaphor.

Rather, its probable absurdity would have been linked to the fact that the words *ideas*[1] and *species* were almost synonymous. Delving into their origins, both had to do with sight. *Idea* is derived from the Greek *idein*, "to see," and shares deeper origins with *vision*. *Species* comes to us from the Late Latin *specere*, "to look at, to see," and that history is echoed in such cognates as *scope* and *spectacle*. Vestiges of those original meanings were still present in the early 1800s; hence "ideas as species" might well have sounded something like "views as perspectives."

The notions of "idea" and "species" were even more tightly coupled in pre-evolutionary science. For millennia, prior to the conceptual convulsion wrought by Darwin and his contemporaries, taxonomists had devoted themselves to making careful, assumed-to-be-stable distinctions among types of living things, oriented by the conviction that every species had essential and unchanging characteristics. That faith in the permanence of species was not at all distant from conceptions of "idea" within Aristotelian metaphysics and Platonic philosophy, by which an Idea/Ideal was understood as an archetype, the pure form towards which a physical entity aspires and tends. A species, that *is*, is literally an Idea within this frame – or, at least, a reflection or instantiation of an Idea – and so not only might "ideas as species" have been heard as redundant by the pre-Darwinian ear; it would likely have sounded as though the terms were reversed.

[1] The terms *idea* and *concept* are used interchangeably in this chapter – in part because the two words are commonly treated as synonyms, but mostly because the etymology of *idea* is useful for framing the discussion.

Given that backdrop, it is not without irony that the device of metaphor is now needed to link the two constructs. Indeed, it is doubly ironic. The meaning of each word has evolved considerably over the past few centuries. For *species*, the rise to dominance of an evolutionary cosmology has occasioned the emergence of a mode of thinking that is as attentive to relationships among forms as to their distinctions. Correspondingly, in the humanities, a century of psychoanalytic, phenomenological, and structuralist thought has redefined *idea* as a contingent, situated, and volatile form that exists in the material world of action and agency, not in some perfect and inaccessible realm. In a strong sense, while the principal meanings of the two words diverged since Darwin offered his redefinition of species, these evolutions have followed weirdly parallel trajectories.

In fact, in light of contemporary discussions of cognition, those trajectories may have re-converged. I explore that possibility in this chapter, prompted in large part by the century-old suggestion from structuralists and phenomenologists that "the learning of an idea/concept" and "the evolution of a species" obey essentially the same processes. More specifically, I argue that there is strong justification for a deliberate effort to cross-fertilize research into the emergence of ideas/concepts and research into the evolution of species.

I draw on three main literatures in this writing – namely those that surround memetics, complexity science, and embodied cognition – with the intention of locating the discussion among the responsibilities of mathematics educators and the interests of mathematics education researchers. As I engage this project, I must be explicit in my conviction that all discussions of this sort are, at their root, explorations in figurative associations. There is no pretence of literal truth in what follows. Rather, fitted to the enterprise of formal education, the discussion is less about unchanging facts and more about utility and provocation.

MEMETICS

Memetics, as proposed by Dawkins (1989), might be defined as a theory of the development and perpetuation of knowledge that is grounded in an analogy between the evolution of concepts and the evolution of life forms.

Of course, that idea is not new (cf. Dewey, 1910). More specifically, memetics aims to do for sociocultural studies what genetics did for biology through positing a base unit of replication and variation: the meme. Drawing an analogy to a gene, Dawkins described a meme as a basic unit of cultural transmission. As he explained:

> just as genes propagate themselves in the gene pool by leaping from body to body via sperms or eggs, so memes propagate themselves in the meme pool by leaping from brain to brain via a process which, in the broad sense, can be called imitation. (p. 192)

Rephrased, a meme is sustained in *mem*ory, and there must be potential for it to be replicated with/in other memories.

Like genes, memes are rarely passed on singly. More typically, they are sustained in large clusters, or *memeplexes*. Memes or memeplexes are thus coherent pieces of information; they are cognitive or behavioural patterns that persist in time. Examples include mythologies, word meanings, habits, skills, fashions, songs, scientific theories, ethical stances, and laws. The word *meme*, in fact, is a meme itself. It has recently been defined in *The Oxford English Dictionary* as an "element of culture that may be considered to be passed on by non-genetic means, esp. imitation."

Clearly the principles of biological evolution, based on physical genes, cannot be directly applied to a conception of cultural evolution that is based on more ethereal memes. For example, genes operate directionally from parent to child; eons have afforded them robust mechanisms for ensuring fidelity of transmission; they take generations to morph significantly. In contrast, memes can spread in multiple directions; they are prone to radical variations as they spread; they can replicate and proliferate at alarming rates; and they can blend and interact with other memes in entirely unpredictable ways.

Nevertheless, some principles of biological evolution have been argued to be directly applicable. For example, Blackmore (1999) asserted the three necessary *and* sufficient aspects of biological evolution – namely, replication, variation, and selection – are also necessary and sufficient within the domain of cultural evolution. In this frame, *replication* refers to mechanisms that enable agents (e.g., genes or memes) to copy themselves from one generation or one host to another; *variation* refers to the potential for difference to arise among agents; and *selection* refers to internal and external conditions that will differentially favour particular variations. Cutting to the chase, memeticists posit that ideas, minds, and cultures emerge as natural selection acts on memes, in a manner that is operationally analogous to natural selection operating on genes. Ideas and species obey the same evolutionary laws.

Some have argued that memes are perhaps more akin to *germs* than *genes* (e.g., Cullen, 1998), based on the conviction that a stronger analogy might be drawn from the behaviours of parasites and infections to the dynamics of meme replication. Indeed, some memes would seem to be almost completely exploitative – that is, viral – including chain letters, Ponzi schemes, computer viruses, urban myths, YouTube videos, and advertising jingles. However, such instances are more the exception than the rule, notable only for their capacity to disrupt more established, relatively stable memes and memeplexes, such as communication systems, the performing arts, monetary systems, political structures, organised sports, literature, science, and mathematics – to name a few.

The "memes are like genes" vs. "memes are like germs" debate serves to highlight one critical assertion of memetics. Each analogy inverts the commonsense assumption that "people have ideas" to the more disruptive "ideas have people." Pragmatically, that is, memetics invites us to think more about how knowings become compelling, how they take hold, how they persist in even highly inhospitable conditions – that is, how ideas possess and exploit knowers, rather than the other way around.

Notably, one of the major critiques of memetics is the usage of such notions as "possess" and "exploit," which can be taken to be suggestive of conscious agency. The same sort of criticism is frequently leveled against Dawkins' (1989) phrase of "selfish gene," and the response in both cases is that such associations are not intended. Rather, these sorts of terms are used to underscore how phenomena that evolve through adaptive processes, such as genes and memes, are coupled to their co-evolving contexts. They are not inert particles interacting in the background, but dynamic influences that enable and constrain possibilities on macro levels of organisation.

As well, the press from a gene-based to a germ-based analogy actually pushes the discussion outside the original bounds of memetics. It is a shift that is tethered to the emergence of a complementary discourse: complexity science. I turn to that topic presently, but pause first to flag the profound similarities of principles of memetics to some of the orthodoxies of Piaget's (2013) genetic epistemology and von Glasersfeld's (1990) radical constructivism. In these and other structuralist epistemologies, learning is seen as rooted in such phenomenological primitives (diSessa, 1982) as placing objects in containers, combining amounts, making distinctions, and arranging shapes. As these primitives – these originary memes – collect, interact, and blend, mathematical insight emerges.

Like Dawkins' memetics, these theories of learning are anchored in evolutionary theory (cf. Dewey, 1910). As compelling and as widely embraced as they are, however, until recently memeticists were virtually silent on the mechanisms by which memeplexes that transcend their meme-parts might arise. That missing element is precisely the focus of complexity science.

COMPLEXITY SCIENCE

Another issue with the construct of *meme* is that it prompts attentions to simplest parts, as though cultural phenomena could be analysed down to their atoms. While that was not Dawkins' intention, his proposal did help steer thinking towards such reductions. Arguably, however, its power has more to do with "emergence." That is, while it is useful to be aware that concepts have constitutive parts, it is vital to be aware how those parts combine and self-transcend. As Kelly (2010) suggested:

> [I]deas never stand alone. They come woven in a web of auxiliary ideas, consequential notions, supporting concepts, foundational assumptions, side effects, and logical consequences and a cascade of subsequent possibilities. Ideas fly in flocks. To hold one idea in mind means to hold a cloud of them. (pp. 44–45)

The suggestion here is more than that memes exist in memeplexes. It is, drawing on complexity science, that memeplexes are a distinct order of being.

Complexity science is the study of emergent systems – that is, collections that somehow self-organise into collectives whose capacities transcend the summed possibilities of their components. Unfortunately, there is no unified definition of *complexity* among researchers. Instead, accounts of complexity research tend to begin by invoking some manner of Aristotle's observation that "the whole can be greater than the sum of the parts," followed by lists that typically include ant colonies (that are more than groups of ants), economies (that are more than collections of consumers), ecosystems (that are more than assemblages of species), and so on. In brief, complex systems researchers investigate both how interacting parts enable systems' global behaviours and how those systems relate to and interact with other phenomena in their environments.

While there is no broadly accepted definition of complex systems, there is near consensus on some identifying qualities. Among those, a complex form is a perceptible coherence that:

- arises in the interactions of multiple agents/subsystems;
- manifests features and capacities that are not observed in those agents/subsystems as it engages with its situation;
- maintains itself over some period of time;
- learns/evolves in response both to internal and to external dynamics.

Provocatively, mathematics can easily be argued to have these qualities. Whereas through much of the history of the western world, the domain has been held up as the epitome of the Ideal (or, at least, the best route to the Ideal, in the Aristotelian sense), today it is not uncommon to encounter suggestions that mathematics "is a living, breathing, changing organism" (Burger & Starbird, 2005, p. xi) or that it "emerges as an autopoietic [i.e., self-creating and self-maintaining] system" (Sfard, 2008, p. 129). More pointedly, Foote (2007) has argued that mathematics is an adaptive, complex system that is approaching the limits of human verifiability. That is, mathematics is a memeplex, a species – one that may be showing signs of exceeding the flesh-based minds that served as the original medium for its emergence.

So, why would agents come together into grander unities in the first place? There are many reasons, but one common impetus is that they are often able to copy themselves more successfully when teamed together. That is, the means by which a memeplex self-perpetuates is also the purpose of that self-perpetuation. There is no external force driving; there is no transcendent form guiding; there is no teleological goal that pulls. Thus, a meme that is not bundled with other memes is much more likely to fade. This point is illustrated in the contrast between *fads* (e.g., pet rocks) that fade as quickly as they appear and *customs* (e.g., Christmas) that are tangled in multiple aspects of culture, such as religion, consumerism, and winter solstice celebrations.

The point is as relevant to the realm of concepts as it is to any other aspect of culture. Consider, for example, the idea of "number." As Barrow (1993) developed, number-as-counting-tool has been around for some time, but for most of its history its utility was limited to discrete applications. It was only recently that "number" was elaborated into a construct that can be used to count, order, measure, locate, rate, and so on. The root of that emergent power is a co-mingling of memes – by which "number as count" has been coupled with "number as measure," "number as a position," "number as movement," "number as value," and so on. Collectively, this flock of memes reaches across a much broader range of situations than the memes do separately – and, importantly, in a manner that gives rise to such greater-than-the-sum-of-their-parts constructs as imaginary numbers. That increased span translates into a greater utility, which in turn increases the likelihood that the knower/host will find value in the memeplex, and thus grow more reliant on that memeplex, which in turn greatly increases the likelihood that any particular meme in that memeplex will be perpetuated.

Therein lies one useful aspect of complexity theory for understanding ideas and concepts. These are never singular or isolated forms. They arise in and are sustained through ecosystems of associations – just as they can be rendered unviable/unthinkable in other ecosystems of associations. Complexity research actually takes the discussion an important step further, as it offers pragmatic advice on how to go about structuring situations that might better support the emergence of complex forms. For example, as developed elsewhere (Davis & Simmt, 2003), the discourse invites attention to redundancy among agents to ensure systemic robustness, diversity among agents to enable adaptive capacity, and interactions among agents to ensure sustainable coherence. These qualities can be manipulated, for those who seek to occasion emergence or to influence the already-emergent.

These qualities are also as relevant to systems of ideas as they are to systems of species. With regard to mathematics and mathematics learning, there is a world of difference between a mechanical system comprising multiple agents/memes/instantiations and a complex system that transcends its agents/memes/instantiations. The latter is better fitted to what is popularly understood as a "concept" within mathematics. But to appreciate how a concept might emerge in the first place, one must look to the medium in which it arose: human knowers.

EMBODIED COGNITION

Alongside a frequent imperative to ensure that learners are exposed to multiple instantiations of concepts, another orthodoxy of contemporary mathematics instruction is that those instantiations should, to the extent possible, be rooted in experience.

Such experiential learning is typically traced back to actions of the physical body – for example, to placing objects in mouths, moving through space, or testing one's reach, which prepare the ground for abstracted understandings of containment, lines, and circles. Perhaps the most detailed example of how such material actions can lead to sophisticated insights is Lakoff and Núñez's (2000) explication of Euler's identity, $e^{\pi i} + 1 = 0$, which they contended begins with the sorts of experiential primitives just mentioned.

As compelling as such arguments are for connecting abstract understanding to physical action, complexity thinking cautions against conflating the original parts with the emergent possibility. That is, while mathematical concepts might be rooted in bodily action, they represent an entirely different sort of entity to those actions – in a manner analogous to the way that human bodies arise in but transcend the co-ordinated activities of cells.

Within a complexity frame, then, the word *embodied* signals something other than physicality; it is a commentary on beingness – which may not be particularly new, given such phrases as "bodies of knowledge," "social corpus," and "body politic." What is novel, however, is that references to bodies are intended to assert existence. The phrase, "body of mathematics," for example, refers to a specific entity.

Phrased differently, one's mathematical competence certainly depends on having a body, but it also depends on that body being a participating element in other bodies. In effect, the notion of embodiment invites consideration of the self-similarity of such phenomena as the development of a personal conceptualisation, the emergence of collective patterns of acting, the historical unfolding of a culture, and the evolution of a species. Understood to exist in dialogical, ecological, and co-emergent relationship, such overlapping and interlacing bodies cannot be understood as distinct – at the same time as they cannot be compressed or conflated.

The import of these realizations arrives early in the trajectories of school mathematics, where there is a necessary elaboration of the notion of embodiment as one nears middle school. Whereas almost all the concepts encountered at the elementary level can be interpreted in terms of (i.e., are analogical to) actions of the physical body – and are typically treated in precisely that way – the analogies for concepts at the secondary level and beyond are mostly mathematical objects (see Hofstadter & Sander, 2013). Whole numbers, for example, are readily interpreted in terms of counts of discrete objects. However, count-based interpretations can frustrate understandings of integers, which are more readily interpreted through such invented instantiations as number lines. Stepping even further, imaginary numbers are only comprehensible through entirely invented objects. While construals of sets, number lines, and rotations can be traced to physical experiences, emergent constructs such as imaginary numbers rely on an entirely different sort of body.

Lakoff and Núñez asserted that a main mechanism here is metaphor, which in effect is a meme that links otherwise conceptually distinct memes. Other devices (also memes) that enable knowers to map experiences from one domain into another include metonymies (Zandieh & Knapp, 2006), gestures (Hall & Nemirovsky, 2012), images (Hoffman, 2000), and logics. Still other devices are crafted into the material world inhabited by the knower (cf. Davis & Sumara, 2005), where the manufactured contexts of existence are not just products of attention and intelligence; they channel awareness and bestow intelligence as well. And while these and other tools can be used as reminders of the role of the body biological in the body of mathematics, they also highlight the role of social bodies. They are tools that, for the most part, arrive from beyond the individual knower, made available through the virtue of being part of a grander body epistemic.

In quite different terms, when the topic of "embodied mathematics" arises in discussions of schooling, it is most often associated with the insight that mathematical understanding emerges from our physical beings. Less often it is used to suggest that humanity embodies mathematics – that is, tautologically, knowledge requires knowers. But there is a third, more radical read of embodiment. Aligned with memetics and complexity, it is that mathematics embodies humans.

Mathematics, that is, requires a corporality, a medium for its maintenance, development, and propagation. Currently that medium is us – where "us" includes languages, symbols, algorithms, digital devices, and all other of our technologies. But a question that emerges at this point is which aspects of "us" may no longer be critical to its sustainability. After all, it has already surpassed the need (and possibility) of human minds to assess its validity. As Foote (2007) noted, more than a decade ago mathematics had reached a place where many of its truths could only be decided by its technological offspring.

This point is a tricky one, as it revolves around an imaginary line drawn between "us" and "beyond us." The crux of the argument is *not* that mathematics has begun to surpass the capacities of the human mind, as this sort of suggestion only makes sense if "mind" is stripped of the tools that amplify its connection-making and idea-maintaining capacities. It is, rather, that humanity is coming to the place where it is compelled to defer this category of decision-making onto a recently emergent order of complex activity. The possibility thus arises of a different sort of embodiment of mathematics, as the mathematics-rooted thinking machine becomes a necessary means of its self-maintenance.

WHAT IS A CONCEPT?

As I prepared to write this chapter, I posed the question, "What is a concept?" to a number of people involved in mathematics education – teachers, undergraduate students, graduate students, colleagues, administrators, and so on. Every response offered could be traced to one of three memes: "concept as tool," "concept as personal construct," and "concept as social construction."

By far the most commonly evoked metaphor was "concept as tool" (also expressed, e.g., as "device" or "instrument of understanding"). A concept, that is, is popularly seen as a pre-formed, static object awaiting a user. Tracking this meme to its practical entailments for educators, if a concept is a tool, learning a concept must be about acquiring that tool – whether by discovering it, assembling it, or being given it. Correspondingly, teaching comes to be about delivering it, presenting it, directing learners' attentions to it, instructing in its construction, and so on.

The second most frequent response, although dwarfed by the first, was some variation of "a social construction." In this frame, there is no "thingness" or "objectivity" to a concept. Rather, a concept is a coherence, a stable vortex of consensual possibility that arises in the medium of social accord – from which it follows that learning is a process of acculturation, acclimatization, induction, or apprenticing, and through which one enters and exploits that space of social accord. Teaching, then, engenders all practices, deliberate and otherwise, to draw another into that space. If spun positively, it is empowering, giving voice, mentoring, consciousness-raising, and inviting participation. If spun negatively, it is more about indoctrinating, subsuming, incorporating.

Clearly, much more can be, should be, and has been said about these and other conceptual flocks of knowing–learning–teaching (see Davis, Sumara, & Luce-Kapler, 2015, for a more thorough treatment). Much less has been offered on possible flock-mates of the metaphor of "ideas/concepts as species," and so I turn to some entailments of that notion for learning and teaching. I do so by relayering memetics, complexity, and embodiment.

Layer 1: From Memetics – A Concept Is a Memeplex

Across the history of knowledge generation, one theme has been consistent: the pace of knowledge production accelerates when meme diversity increases. Hence the many strategies for placing many minds in closer proximity, such as cities, schools, skyscrapers, high-speed travel, increased longevity, and the internet. This detail has not been lost on mathematics educators. Indeed, Dawkins' suggestion of memes is highly reflective of mathematics educators' focus on multiple instantiations/representations.

There is one important difference. Memetics posits that the meme, within its own frame, has no other purpose but self-replication. Memes that are stickier, that propagate more easily, or that offer great utility have a greater chance of replicating in host minds. Learning, it follows, is less about assembling memes and more about allowing memes to find one another – a point that helps us understand how, given the incredible human capacity to construe relationships among the unrelated, we generally seem to make the same sorts of linkages around mathematics concepts. Some memes exhibit a sort of preferential attachment for one another. A mathematical concept would appear to be such a beast – that is, a memeplex comprising memes with high affinities.

To be clear, this interpretation diminishes the agency of the learner in learning. Simply put, we do not have much control over the ideas that overtake us, or the manner in which they self-assemble. While these realisations offer little direct advice for formatting teaching, they do lend credence to the current orthodoxy of presenting multiple representations for concepts. They also validate the current concern for teachers' disciplinary knowledge of mathematics. Teachers with more robust memeplexes are likely to offer richer, more nuanced, and better-articulated meme sets for learners.

Layer 2: From Memetics + Complexity Science – A Concept Is a Life Form

It is one thing to assert that self-sustaining memes serve as the bases of concepts, and quite another to suggest that a concept is a vibrant entity. Or, in more graphic terms, a concept is a life form, one that lives in/on/across the minds of knowers – and one that not only has the meme's capacity to hijack resources of other entities to replicate, but also that can self-maintain, grow, and enter into mutually supportive relationships with other similar forms.

So framed, concepts are more than emergent memeplexes. They are agents interacting – that is, collaborating and competing – with other agents in ecosystems. Like other life forms, they are survival-oriented. They gravitate toward, more hospitable environments; they flock with compatible concepts; they find ways to become necessary to their host.

"Learning" here is indistinguishable from "evolution." Which of these terms the observer chooses to attach to observed transformations depends on that observer's assumptions about agency and causality – or, more metaphorically, whether the observer perceives the developing concept as an emerging idea or an emerging species.

As for formal education, teaching becomes a participation in the propagation of the concept. In the extreme, it is an enslavement of sorts, as it entails being sufficiently held by species/ideas/concepts to be unable to do otherwise than seek their perpetuation through helping create hospitable homes in other minds.

Layer 3: From Memetics + Complexity Science + Embodied Cognition – A Concept Is a Living Body

One of the interesting things about the etymologies of words such as *body* and *corpus* is that their deep origins have been lost. Traces typically end abruptly with phrases like "unknown origin" or "meaning uncertain."

One explanation for these dead ends is the simple physicality of the human body. It is what it is, a literal form and not a figurative extension of something else. By contrast, the etymologies of most words are genealogies of metaphors that most often reach back to some fundamental bodily action or body part.

But there is something unsatisfying with this explanation, made evident in such phrases as "body of knowledge" and "social corpus." In these cases, the body is a constellation of relationships. It lives and thrives – and while it is dependent on "lower" (in the complexity sense of more partial, not as lesser ranked) physical bodies, it is not reducible to them.

What is interesting, and perhaps unique, about concepts is that it is not hard to imagine that they might evolve to a point of no longer needing their biological hosts. This thought has been a favorite of futurists and science-fiction writers. With exponential increases in technological possibility, more and more modern seers find merit in the idea that digital and logical, silicon-based life forms might soon succeed their analog and analogical, carbon-based progenitors.

It is not my purpose here to add to these speculations. However, such predictions are germane to the topic at hand. Their increasingly compelling nature affords some insight into emergent beliefs about "concepts" – and, in particular, to the plausibility of the suggestion that *ideas are species* – literally, and not metaphorically. It has become easy to believe that our conceptual successes may end up being our successors.

AND SO ...?

I taught Grade 8 mathematics through the 1980s. Among my great frustrations was the task of introducing operations on integers, especially multiplication, in meaningful ways. Inevitably, when the topic arose, students tossed out an array of memes that were not particularly useful for a vibrant and robust memeplex: "two wrongs make a right"; "same sign, positive; different sign, negative"; etc.

In retrospect, what is abundantly clear is that "concepts" in the body of school mathematics at that time were not evolving, networked coherences but static, isolated assemblages. Consequently, there was little hope of developing mathematically powerful alternatives to the memes that had entrenched themselves. Simple mnemonics and dead-end rules choked out the very possibility of images and metaphors that might constitute farther-reaching, self-maintaining ecosystems of associations.

For me, the inverted suggestion that "ideas hold people" rather than "people hold ideas" is particularly useful for making sense of the educator's responsibilities here. It speaks to a particular kind of knowing – namely, an obligation to be held by the sorts of memes that are generative rather than merely utilitarian. In particular, an embrace of concepts as living forms compels a very different attitude towards such matters as teacher disciplinary knowledge. For the educator, it is not enough to have "acquired knowledge" or to have "developed mastery"; rather, one must surrender one's mind as a medium for the propagation of ideas.

To that end, the meta-memeplex that collects memetics, complexity thinking, and embodiment would seem particularly fecund – even if not especially well suited to the current ecosystem of educational thought. For the past century, efforts to evolve formal education have been frustrated by the teeter-totter debate between advocates of standardised education and advocates of authentic education. The former continue to prevail, buoyed by a flock of memes that include the ideals of objective knowledge, standardised outputs, and efficient productivity. Arguably, no other model of formal education is so well fitted to the scientised and commercialised culture in which schools must operate. On the other end of the see-saw, authentic education aligns with cultural ideals of personalisation and diversity, along with increasingly compelling literatures on learning out of neurology, psychology, sociology, and anthropology.

Observed from the outside, and as evidenced by wave after failed wave of educational reform, it is obvious that both camps are held by their concepts, not the other way around. Against that backdrop, it would seem that a rethinking of "concept" in the terms presented here would invite nuanced conversations on the flocks of figurative associations that render some ideas compelling and others nonsensical. In fact, it could be seen as an ethical obligation. In a domain devoted to the preservation and development of insight, it might be argued that no other stance is defensible.

REFERENCES

Barrow, J. D. (1993). *Pi in the sky: Counting, thinking, and being*. New York: Little, Brown and Company.

Blackmore, S. J. (1999).*The meme machine*. Oxford: Oxford University Press.

Burger, E. B., & Starbird, M. (2005). *The heart of mathematics: An invitation to effective thinking* (2nd ed.). Emeryville, CA: Key College Publishing.

Cullen, B. (1998). Parasite ecology and the evolution of religion. In F. Heylighen, J. Bollen, & A. Riegler (Eds.), *The evolution of complexity* (pp. 255–261). Dordrecht, The Netherlands: Kluwer Academic Publishers.

Davis, B., & Simmt, E. (2003). Understanding learning systems: Mathematics teaching and complexity science. *Journal for Research in Mathematics Education*, 34(2), 137–167.
Davis, B., & Sumara, D. (2005). Challenging images of knowing: Complexity science and educational research. *International Journal of Qualitative Studies in Education*, 18(3), 305–321.
Davis, B., Sumara, D., & Luce-Kapler, R. (2015). *Engaging minds: Cultures of education and practices of teaching*. New York: Routledge.
Dawkins, R. (1989). *The selfish gene* (2nd ed.). Oxford: Oxford University Press.
Dewey, J. (1910). *The influence of Darwin on philosophy*. New York: Henry Holt.
diSessa, A. A. (1982). Unlearning Aristotelian physics: A study of knowledge-based learning. *Cognitive Science*, 6(1), 37–75.
Foote, R. (2007). Mathematics and complex systems. *Science*, 318(5849), 410–412.
Hall, R., & Nemirovsky, R. (2012). Introduction to the special issue: Modalities of body engagement in mathematical activity and learning. *Journal of the Learning Sciences*, 21(2), 207–215.
Hoffman, D. (2000). *Visual intelligence: How we create what we see*. New York: W. W. Norton.
Hofstadter, D., & Sander, E. (2013). *Surfaces and essences: Analogy as the fuel and fire of thinking*. New York: Basic Books.
Kelly, K. (2010). *What technology wants*. New York: Penguin.
Lakoff, G., & Núñez, R. (2000). *Where mathematics comes from: How the embodied mind brings mathematics into being*. New York: Basic Books.
Maturana, H., & Varela, F. (1979). *Autopoiesis and cognition: The realization of the living*. Boston, MA: D. Reidel.
Mitchell, M. (2009). *Complexity: A guided tour*. Oxford: Oxford University Press.
Piaget, J. (2013). *Principles of genetic epistemology: Selected works* (vol. 7). London: Routledge.
Sfard, A. (2008). *Thinking as communicating: Human development, the growth of discourses, and mathematizing*. New York: Cambridge University Press.
von Glasersfeld, E. (1990). An exposition of constructivism: Why some like it radical. In R. B. Davis, C. A. Maher, & N. Noddings (Eds.), *Constructivist views on the teaching and learning of mathematics* (pp. 19–29). Reston, VA: National Council of Teachers of Mathematics.
Zandieh, M. J., & Knapp, J. (2006). Exploring the role of metonymy in mathematical understanding and reasoning: The concept of derivative as an example. *The Journal of Mathematical Behavior*, 25(1), 1–17.

15

Inhabiting Mathematical Concepts

RICARDO NEMIROVSKY

Elucidating the concept of concept looms as a self-referential effort, one of those that are typically drowned in mind-boggling paradoxes. There are no concept-less anchors or grounds to stand on, from which the nature of concepts arises to be contemplated. My opening effort will be to discern images of the conceptual, as styles of thoughts, assumptions, language games, expectations, stereotypes, and so forth – very much in the ways we differentiate pictorial or literary styles. First, I attempt to characterise what I call the 'classic' image. After navigating criticisms of the classic image, I turn towards an alternative image, which I call the 'crystalline image' of concepts. This crystalline image will surround the ensuing exploration of the mathematical concept of number.

THE CLASSIC IMAGE

Aristotle is the larger-than-life name for the classic image of the concept. This image entails a tree-diagram, such that branches bifurcate into other branches (Figure 15.1). Each branch bifurcation radiates from a common node located on a thicker branch. Branches extend towards other nodes originating thinner branches. Entities – anything that *is* – are located in proper nodes. For instance, human beings are located in a node derived from 'animal' and bifurcated towards 'rational.' Each node encompasses a class of entities defined by a chain of genre-species. A class is constituted by 'blocking' further bifurcations underneath; for example, the class of human being is constituted by brushing aside such bifurcations as 'male/female,' 'adult/child' and infinitely many others. Each class correlates to a 'concept', such that the web of predicates proper to the entities of the class outlines its concept. The arboreal diagram seems largely to emanate detachedly from the things it organises, such as the categories of

FIGURE 15.1 A Tree of Classes and Concepts.

'rationality' and 'animality' as given *a priori*, in order properly to define one of us as a human. Not all differences are of equal relevance, of course; some are essential while others are accidental. Essential differences, such as rationality, discriminate natural kinds. They are to be sharply distinguished from contingent traits, everywhere misleading us into assuming pseudo-classes associated with pseudo-concepts. This arboreal image expands into central taxonomic regions over which classes or categories appear clearly located; its upper and lower edges, however, tend to blur and falter. On the upper side, higher genera correspond to larger classes with fewer traits in common. Highest of all, there is the absolute class – a category for all that is, a universe without noticeable regions, beyond grasp and discernment. On the lower side, accidental differences group entities without order and principles so that senseless gatherings are obtained instead of categorical classes.

I will highlight two criticisms of the classic image. The first one is that it tends to reduce the becoming of entities to a matter of opaque jumps between contiguous classes and/or successive stages. Classes appear fully constituted in the midst of other classes equally completed. Even if the branches were created on the basis of lineage or ancestry, so that lower, more recent species are to be derived from upper, older ones, they depict a temporal evolution in terms of stages, whose becoming gets compressed onto the vertical single-event steps of a staircase. A second criticism is that essential differences are, at least in many realms of life, stereotypes imposed by cultural and political enforcements; not a matter of natural kinds, but of taxonomies driven by relations of power and exclusion, as a result of which the interstitial spaces between classes overflow with that which is unfit for properly constituted classes. The appearance of the unfit falling off the grid, such as a person who is neither male nor female, or someone who is neither native nor alien, is typically associated with the troubling appearance of monstrosities and unsightly cases (e.g. a bearded woman, a gigantic ant).

THE CRYSTALLINE IMAGE

The alternative to the classic image that I pursue is inspired by processes of growth, decay, and individuation, such as the ones of crystal growth, which motivates the name of the image advanced here, or the genesis of a geological formation. These images depict concepts not as webs of predicates proper to formed classes, but themselves as always under formation, growing and decaying in the intercourse of various unpredictable flows (Figure 15.2). According to the crystalline image, concepts offer transient shelters and places, shaping and shaped by all the varieties of life inhabiting them, such as birds and worms. Inhabiting a concept entails roughly going along – doing, making, and perceiving – meshes of trails carved out over periods spanning natural and social history. In order to describe this crystalline process of carving out and appearance of the conceptual, I briefly elaborate on virtuals, affects, and senses.

Virtuals

Nucleation is a first phase of crystal growth involving the formation of particles immersed in a supersaturated or supercooled medium. If these particles become larger than a critical size, they may grow into crystals:

> Particles smaller than a critical size, which may be referred to as clusters, are formed in the system. Some of these clusters, out of the many that repeatedly come together and part again, may by chance grow larger than the critical size. (Sunagawa, 2005, locations 375–376)

Most commonly, clusters get formed around either impurity components or irregularities on the surface wall of a container. Characterizing a certain cluster by its chemical composition and shape is informative in some regards, but it is not enough to illuminate what it may grow or decay

FIGURE 15.2 A Salt Crystal Immersed in Flows of Matter and Heat.

into. A cluster may grow into a crystal, partition, dissolve, become part of another cluster, and so on, indefinitely. I call the world of what a cluster may become its virtual or virtuality. Virtuals extend much beyond the 'intrinsic' traits, such as the chemical composition of a cluster at a given time, because they are relational. What counts for a virtual are relations of all kinds; a cluster, for instance, becomes in relations of proximity, of flow and viscosity, of temperature and density distributions, and of distant electromagnetic fields, among others. Virtuals are not amenable to full determination because the relations that count for them are open-ended, not yet set, and underspecified.

Deleuze (1994) distinguishes between the virtual and the possible. I understand the possible of an entity as a distribution of classes or categories that the entity may become a member of. For example, someone who is familiar with a child may consider the likelihood for the child to grow into a lawyer, monk, politician, or wanderer. Criteria for the possible assume the existence of certain stable classes with membership requirements and procedures that, in this example, may appear to be more or less suited to the character of the child. The virtuality of the child, on the other hand, is not to be mapped out by a cartography of distinct paths ending in established destinations, not only because these destinations are themselves in flux, but also because, instead of 'final' destinations, the virtual is pervaded by never ending processes of individuation. Even if the child were to become a monk, and so realise a certain likelihood for that possibility, the virtual surpasses class membership and interrogates what life as a monk he is going to lead, in what ways he is to become a unique monk distinguishable from all the other monks, and stays always open to his eventual drifting into other-than-monk individuation. The virtual is delimited not by defined boundaries, but as a horizon: as we explore a virtuality, the horizon recedes, opening up hitherto-occluded regions.

Living beings carry on their lives by dealing with virtuals. Choosing a piece of food is driven by a sense of its virtuality, such as its upcoming to be more or less tasty, digestible, or salutary.

Affects

Since I characterise affects as multiplicities of interpretations and feelings, in this subsection I elaborate on the notions of interpretation and feeling and their mutual relationships. I think that an *interpretation* is closely akin to what Wittgenstein called 'seeing as': now I see this drawing as a box, now I see it as a polyhedron, now as a drum – each is an interpretation. In fact, every perception is an interpretation, although usually we only realise this when a perception transforms itself into something else (e.g. box -> drum). That all living beings perceive entails that all of them interpret.

I start my commentary on *feeling* by noting that every perception is embedded in feelings, which also means that every perception awakens aesthetic qualities. Buck-Morss (1992) reminds us of the etymology of 'aesthetics':

> *Aisthitikos* is the ancient Greek word for that which is 'perceptive by feeling.' *Aisthisis* is the sensory experience of perception. The original field of aesthetics is not art but reality – corporeal, material nature. As Terry Eagleton writes: 'Aesthetics is born as a discourse of the body.' It is a form of cognition, achieved through … the whole corporeal sensorium. (p. 6)

The body emanates feelings permeating that which is perceived with aesthetic qualities: this room is somber, that dress is funny, etc. A living body lets things, others, and symbols acquire felt qualities.

There is a very close interplay between *feeling* and *interpretation*. I see the bus I want to ride coming. It is frustrating, because I am too far from the bus stop to get there in time. When the bus gets closer, I notice that it is not the one I want to ride; now I feel relieved not to be missing the one I am waiting for. And so on. Changes of interpretation (my bus, not my bus) mesh with changes of feeling (Oh, too late! Well, it's fine) and vice versa. Feeling and interpretation are not external to each other. Their differences are creative, not oppositional. Oppositional differences are germane to the ordinary image of difference between two entities – she and her companion, a car and a motorcycle – with common traits that make them comparable and uncommon traits indicating their opposition and mutual negation. Deleuze worked to develop an alternative philosophy of difference, centered on the idea of difference-in-itself or intrinsic difference. Difference-in-itself is not one of opposition but of genesis and creativity. Feelings and interpretations displace and stimulate each other, prompting the circulation of affects.

Senses

I use the word 'sense' as it is meant in certain regular English utterances, such as 'Give me a *sense* of the work that you do,' or 'In what *sense* do you say that you are aggressive?' This sense of *sense* has received great philosophical development after Frege's (1892/1980) introduction of the difference between sense and reference, according to which 'the morning star' and 'the evening star', for instance, connote two different senses for the planet Venus, which is their common reference. In *The Logic of Sense*, Deleuze (1990) articulates the notion that a sense is infinite, since no finite list of determinations will suffice to articulate it fully. Usually, a speaker conveys a sense by means of examples and suggestive allusions that are intended to awaken in the listener a boundless background of feelings and interpretations, as I just did it in the first sentence of this paragraph.

As affects circulate through and by living beings, certain senses get traced and delineated. For instance, the sense that a certain neighborhood is safe, or that it is dangerous, is carved out by flows of affects permeating those who pass through or live in that neighborhood. The infinity of a sense entails that it is not derived from definitions, but from how it is inhabited. The sense that a neighborhood is safe, for instance, is expressed by how people inhabit it, or the sense that a movie is fascinating is to be recognised by how some viewers get fascinated by it. Flowing affects shape senses not unlike how a canyon is crafted over millennia by the shifting flows of a river. For example, a sense that public schooling is a precious social achievement to be defended and nurtured permeates numberless institutional commitments, political allegiances, historical events, and daily actions undertaken by millions of families. Like a geological formation, this sense of public education is subject to change as other affects begin to circulate, leading to erosion and gradual molding of something different. A sense grows or decays as some affects spread and proliferate, while others fade. A sense withers because carving affects cease to circulate, as a result of which entropy sets in: dust and erosion make these senses vanish and go underground, so that archeological work becomes necessary to retrieve some of their traces. Because the roots of a sense are affects flowing through communities and institutions, a sense is to be mapped over geographies.

Returning to my prior statement, 'Living beings carry on their lives by dealing with virtuals,' we can now grasp that, like footprints on muddy ground, virtuals are impressed by senses. For example, the sense a teacher undergoes for a certain student's ability to excel in mathematics impresses the student's virtuality as a mathematician. As her teacher inhabits a sense of, say, her great capacity to become a mathematician, the teacher may stimulate the student to participate in certain events or to explore a topic further, and in these ways footprint the student's virtuality as a mathematician. This kind of footprinting is never deterministic, because flowing affects may prompt the student to become other than a mathematician, or the way in which the teacher enacts her sense may turn the student in opposite directions, but the general point is that the senses of a virtual are part and parcel of that virtual. For this reason, the prepositions 'of' and 'for' both befit the meshing of virtuals and senses: a sense is of/for a virtual.

The upshot of this detour through virtuals, affects, and senses is that it enables me to clear up a crystalline image of concepts: a concept is an assemblage of all the senses of/for a virtual. Whereas in the classical image a concept gathers all the predicates proper to a class, in the crystalline image a concept meshes all the senses of a virtual, so that the layout of a concept is unceasingly being forged by the shifting senses of a virtual. According to the classic image, people know or have concepts to the extent that they generate appropriate predicates; according to the crystalline image, groups, communities, and individuals – human and other than human – inhabit senses flowing along geographical regions of a concept. Individuals do not 'have' a concept; they inhabit some of its regions, making these, by their ways of life, distinctive and second nature.

THE CONCEPT OF NUMBER: FORGING NEW SENSES FOR THE REALITY OF NUMBERS

The concept of number is and has always been in flux, gathering number senses of the most diverse kinds, from subitizing small numbers by birds and babies to ascertaining numberings for infinite sets. Rather than trying to make sense of the immense virtuality of number, which amounts to pursuing a fast-receding horizon, I focus on particular events that took place during the 1880s in response to the 1883 publication of the *Grundlagen* (Cantor, 1883/1976). First, I describe ideas articulated in this paper, and then trace how it sparked off flows of affects forging new senses for the reality of numbers and for the nature of infinitesimals.

Cantor's *Grundlagen*

In 1883, Cantor published the *Grundlagen*, in which he introduced transfinite ordinal numbers. Cantor proposed two principles of generation for numbers. The first one is the successive addition of units (e.g. 3+1 -> 4), the second one is applicable to a succession of numbers for which there is no largest one; by means of this second principle, a next number gets generated that is larger than all of them. The sequence of natural numbers has no largest number; therefore, using the second principle of generation, we can define a next number that Cantor represented with the letter 'ω':

$$1, 2, 3, \ldots, \omega$$

Because every counting step takes some time to complete, if we were counting from any natural number on, we would never get in finite time to utter 'ω.' However, we can attain a number without counting the whole sequence; for instance, 7 can be reached just by asserting that it comes after 6. The same is true for ω: it comes right after *all* the natural numbers. Furthermore, ω is a number that comes after not only the sequence of natural numbers but also after any sequence of numbers such that an ordered one-to-one correspondence can be established with the natural numbers. For example, the sequence of even numbers can be set in a one-to-one correspondence with the natural numbers in a way that maintains relations of order between them, so that, given two terms corresponding to each other, their successors correspond to each other as well:

1	2
2	4
3	6
4	8
...	...

Therefore, both sequences are followed by the same ω. This is also expressed by saying that a transfinite ordinal number corresponds to an *order type*.

Having defined ω, by applying the first principle we obtain:

$$1, 2, 3, \ldots, \omega, \omega{+}1, \omega{+}2 \ldots$$

this sequence has no largest number, therefore the second principle generates:

$$1, 2, 3, \ldots, \omega, \omega{+}1, \omega{+}2 \ldots \omega + \omega, \ldots \qquad (1)$$

Given that ω + ω = 2ω and applying again the first principle:

$$1, 2, 3, \ldots, \omega, \omega{+}1, \omega{+}2 \ldots. 2\omega, 2\omega{+}1, 2\omega{+}2, \ldots$$

and so on.

The *Grundlagen* showed that the operations of addition and multiplication, among others, are applicable to the ω's. Their properties can be different though. For example, in:

$$1 + \omega \neq \omega + 1$$

the terms 1 + ω and ω + 1 correspond to different order types because a sequence for the left side has no largest number, whereas one for the right side has:

$$\omega + 1 \Leftrightarrow 1, 2, 3, 4, \ldots. \omega{+}1$$

Each term in the sequence (1) of transfinite ordinal numbers, starting with ω, corresponds to a set that has the cardinality of the natural numbers. Two sets have the same cardinality if a one-to-one correspondence, not necessarily order-preserving, can be defined between their elements. If we disregarded order, a sequence for ω (e.g. 2, 4, 6 …) and another for ω + ω (e.g. 1, 2, 3, … ω, ω + 1, ω + 2, ω + 3 …), for example, can both be set into a one–to-one correspondence with the natural numbers:

2	1	↔	1
4	ω + 1	↔	2
6	2	↔	3
8	ω + 2	↔	4
10	3	↔	5
12	ω + 3	↔	6
….	…	↔	…

This result was perplexing because, in contrast to the natural numbers, for which each cardinality corresponds to only one ordinality (e.g. 4 ⇔ 4th), in the case of the transfinite numbers an infinity of ordinals all appeared to correspond to the same cardinality.

The publication of the *Grundlagen*, in which for the first time Cantor used the term 'number' in the context of transfinite arithmetic and a symbol for the infinite different from ∞, awakened a wide range of affects. Some of these originated from long-held perceptions of the infinite as an unreliable source of contradictions to be eschewed in mathematics, motivating a general rejection of transfinite numbers. Many other affects focused on certain aspects of Cantor's ideas; of these, I briefly trace two threads of affects stirred by the *Grundlagen*: (1) the reality of numbers, and (2) the debate about infinitesimals.

Thread 1: The Reality of Numbers

In his *Grundlagen*, Cantor associated the reality of the concept of number with its appearance in inner intuition: 'The concept of number of elements has an immediate objective representation in our inner intuition' (Cantor, 1883/1976, p. 71). He also described a process of concept formation as creating the conditions 'for the awakening of the concept, which has slumbered within us' (p. 95). He articulated these rather enigmatic remarks more explicitly over the following years. In a paper published in 1887, he illustrated the origins of the concept of number as follows:

> For the formation of the general concept 'five' one needs *only* a set (for example all the fingers of my right hand) which corresponds to this cardinal number; the act of abstraction with respect to both the properties and the order in which I encounter these wholly distinct things, produces or rather awakens the concept 'five' in my mind. (Cantor, quoted in Ortiz Hill, 2003, p. 112)

The objective intuition of number, then, is obtained through an act of abstraction that ignores all properties of a set except for how many elements it has. Cantor proposed that acts of abstraction are the conceptual source of all numbers, including finite and transfinite numbers. This approach was highly praised by Husserl who was, at the time, developing a cognate theory of the number concept. In 1892, Frege published a review of Cantor's 1887 paper in which he criticized grounding the concept of number on abstraction because 'it presupposes that we are already acquainted with what is to be defined' and because 'the verb 'abstract' is a psychological expression and, as such, to be avoided in mathematics' (Frege, 1892/1994, pp. 180–181). Frege accused Cantor of 'psychologism' – a type of empiricism incompatible with the logical rigor and apodictic nature of mathematics. Psychologism deviated from what came to be known as 'logicism,' an approach championed by Frege. Many predicates, according to Frege, articulate concepts, so that, for instance, the statement 'horses are quadrupeds' conveys the concept 'quadruped' and asserts that horses fall under that concept. Perceptions and intuitions are inimical to deriving inferences from concepts and objects that fall under them. Frege developed a definition of number in terms of extensions of a second-order concept (i.e. concept of a concept). Psychologism and logicism name complex and multifarious flows of affects that embroiled practitioners of philosophy, science, mathematics, theology, law, and others: all the affects that swirled through interpretations of truth and certainty.

Thread 2: The Debate about Infinitesimals

The *Grundlagen* elaborates on the difference between the genuine and improper infinites. The improper or potential infinite is an infinite that can be approached but not reached. The defining characteristic of the potential infinite is the use of finite magnitudes to be arbitrarily close to it. Genuine infinite entails the actual obtainment of infinite magnitudes. A major aim of the *Grundlagen* was the acceptance of the genuine infinite and the dismissal of the 'fear and wholesome terror' it often inspired by being identified as a mathematically dark place in which 'all is possible' (Cantor, 1883/1976, p. 74).

While this is the *Grundlagen*'s stance regarding the infinitely large, Cantor rejected the genuine infinite for the infinitely small:

> All attempts to transform the infinitely small by force into a genuinely infinitely small magnitude should be abandoned as purposeless. (p. 74)

This entailed a dismissal of infinitesimals – numbers greater than zero and less than any real number. While other mathematicians proposed the reciprocal of a transfinite ordinal number (e.g. $1/\omega$) as an infinitesimal number, Cantor called them 'paper numbers.' A few years later, Cantor attempted to prove that infinitesimals were mathematically untenable by showing that they are inconsistent with the Axiom of Archimedes (i.e. 'given real numbers $a < b$, it is possible, for some arbitrarily large $n \in N$, to ensure that $na > b$'). In order to strengthen such an interpretation, Cantor worked to show that the Axiom of Archimedes is not, in fact, an axiom, but a theorem about real numbers. This was important because an axiom can be taken as an arbitrary choice, whereas a theorem cannot be contradicted without assailing the system of real numbers.

REFLECTIONS

In focusing on the contribution of Cantor, one might be inclined to see him as an author. In this reflection, I elaborate on what, precisely, Cantor was an author of. Cantor's attributed authorship for the *Grundlagen* intertwined and co-moved his biography, his mesh of social relationships, his academic career, and his theological, philosophical, and mathematical orientations. I interrogate the nature of such authorship in relation to the drift and growth of the mathematical concept of number. In order to explore this, I note some aspects integral to crystal growth.

Singularities

Unless crystals are grown in extremely controlled circumstances, they are pervaded by 'defects' or 'imperfections' breaking the regularity of their lattices. Technologies for the growth of defect-free crystals are known only for some single component ones, such as silicon. There are different types of imperfections: some are point-wise, such as single atomic impurities or vacancies; others are lattice-wide, such as the linear fractures called 'dislocations.' Material scientists started to learn in the 1920s that imperfections are crucial to crystal growth. Under wide ranges of temperature and saturation levels, without imperfections, crystal growth slows to a standstill. Furthermore, for the most part, imperfections are the sites of growth, in the sense that the growth of a crystal radiates from them. They are also sites of contingency because their unforeseeable and local circumstances project onto the traits of the lattices growing around them. Impurities and dislocations mould and prompt the flows of heat and mass necessary for the occurrence of crystal growth and dissolution.

By using the term 'singularity' in reference to imperfections, impurities, dislocations, and so forth, as applicable to the processes of conceptual growth, I propose that the *Grundlagen* was a singularity for the growth and dissolution of the number concept. The more general point is that Cantor authored a singularity, not a concept. Cantor's text was a local disruption awakening and perturbing powerful flows of affects, which traversed the work of many mathematicians and conferred shape – like rivers and winds carving a geological formation – to what came to be set theory and the new axiomatic foundations of early twentieth-century mathematics.

Matters

The *Grundlagen* constituted a singularity for the growth and dissolution of not only the mathematical concept of number. The text included significant reflections of a theological–philosophical nature and elicited correspondence with scholars in these fields. However, focusing on how it moulded and prompted the circulation of mathematical affects, this question arises: How are mathematical concepts distinctive? How are mathematical concepts different from, say, musical or medical concepts? What made the *Grundlagen* a dislocation in mathematics and not in sociology?

My first thought is that concepts are rarely 'of' a single discipline. There are many senses of number, for example, whose mathematical-ness is, at least from some strict viewpoints, debatable (e.g. What makes certain numbers lucky?). Perhaps what we usually call 'technical terms' refer to concepts that are more exclusive of a single discipline: 'technical' implies that it is to be used by and for technicians. In topology and differential geometry 'manifold' is a technical term. However, if we examined not its etymology but the mathematical history of 'manifold,' we readily encounter that it is distinct, and yet deeply connected to other concepts, such as surface, sub-space, and local flatness, whose boundaries with other-than-mathematics are elusive, if not impossible, to demarcate.

Returning to the idea that concepts gather all the senses of a virtual, many of the senses for the virtuality of number are unambiguously mathematical, but not all of them. The question needs to be reformulated then: What makes some of the senses gathered by a concept specifically mathematical? I would say that it is the way they intermingle with other senses. Mathematical-ness is a mesh of senses. For instance, many number senses intermingle with various senses of/for the infinite, the continuum, the line, and so forth. Then the challenge of recognizing mathematical-ness is to get a sense for a mesh of senses. It comes down to Wittgenstein's ideas about concepts in terms of thin and wavy threads meshing into a rope. We can study what the threads are 'made of,' how they are strung together, how long they are, the range of weights they safely hold, how easy is to bend them without breaking them apart, what dyes will colour them, and so forth. These are the *matters* of a rope or fabric: those aspects that matter for it. As with any other fabric of senses, mathematical-ness has its own matters. The matters of a fabric are neither clear-cut nor universal; something in a fabric that is the matter for an upholsterer might not be so for a curtain maker. And the same is the case in mathematics. Something that is the matter for a nurse to set up a prescribed dose might not be so for a programmer computing newer pi's digits. I think there are no bases to determine the matters of mathematics other than choosing 'wisely' a few of them to portray.

A mathematical matter that permeated the conceptual growth stimulated by the *Grundlagen* is the infinite. The tension finite/infinite is momentous for number senses and threads in mathematics across time and geographies. Cantor's use of a symbol for the infinite different from ∞ attempted to express a turning point in the history of the infinite, which, the latter being a mathematical matter, implied also a turning point in the history of mathematics.

Beyond Invention vs Discovery

A recurrent theme for philosophy of mathematics is the extent to which mathematics is invented or discovered. It seems pertinent to ascertain whether the authorship of a singularity, such as the *Grundlagen*, amounts to invention or discovery. However, I think that in an important sense a concept is neither an invention nor a discovery: it emerges in the course of social life with neither the planning intent characteristic of an invention, nor the conscious realization of what had been there all along, as a discovery entails. By tracing two threads of affects prompted by the *Grundlagen*, I have tried to illustrate the thesis that individual or group authoring can be correlated to the production of a singularity, not of a concept. Concepts emerge similarly to how the colourations and textures of a path frequently walked along gradually appear. Ingold (2015, p. 61) illustrates how a path grows, incorporating traces of life:

> [In] northern Namibia, … indigenous Akhoe Hai//om hunter-gatherers, according to their ethnographer Thomas Widlok, have unwittingly created paths through the desert, primarily between water-pans, in the form of lines of mangetti trees. As they went on their way, people would chew the highly prized nuts of these trees, periodically spitting out the hard kernels from which new trees grew. And although the trees have a short life-span, once the path is made it is conducive to further use as the trees provide food in the form of nuts, shade from the hot sun and water that collects in the hollows of old trunks.

The paths created by the Akhoe Hai//om are neither invented nor discovered; rather, the paths emerge from their forms of life and the geographies offering sanctuary to them. Similarly, transfinite mathematics emerged through the wanderings of many mathematicians and institutions, for whom the *Grundlagen* was a fruitful tree among others.

Hilbert famously wrote, 'No one will drive us from the paradise which Cantor created for us.' Regardless of the fact that what was a paradise for Hilbert was an inferno for some, the important point, I think, is that concepts are vast dwellings, landscapes to inhabit by virtue of their being amenable to hosting certain forms of life and not others, while, at the same time, those who inhabit a concept transform it as they go on living in it.

ACKNOWLEDGEMENTS

Writing of this paper was supported in part by the InforMath project, funded by the National Science Foundation through grant DRL-1323587. All opinions and analysis expressed herein are those of the author and do not necessarily represent the position or policies of the Foundation. I am deeply grateful to Elizabeth de Freitas for the feedback she offered on a draft of this chapter, and for the inspiring conversations we have had on multiple occasions.

REFERENCES

Buck-Morss, S. (1992). Aesthetics and anaesthetics: Walter Benjamin's artwork essay reconsidered. *October*, 62, 3–41.

Cantor, G. (1883/1976). Foundations of a general theory of manifolds. *Campaigner*, 9(1–2), 69–97.

Deleuze, G. (1990). *The logic of sense*. New York: Columbia University Press.

(1994). *Difference and repetition*. New York: Columbia University Press.

Frege, G. (1892/1980). On sense and reference. In P. Geach & M. Black (Eds.), *Translations from the philosophical writings of Gottlob Frege* (pp. 56–78). Oxford: Blackwell.

(1892/1984). Review: Cantor, theory of the transfinite. In B. McGuiness (Ed.), *Gottlob Frege: Collected papers on mathematics, logic, and philosophy* (pp. 178–181). Oxford: Blackwell.

Ingold, T. (2015). *The Life of Lines*. London: Routledge.

Ortiz Hill, C. (2003). Abstraction and idealization in Edmund Husserl and Georg Cantor prior to 1895. In C. Ortiz Hill & G. E. R. Haddock (Eds.), *Husserl or Frege? Meaning, objectivity, and mathematics* (pp. 110–135). Chicago, IL: Open Court.

Sunagawa, I. (2005). *Crystals: Growth, morphology, and perfection*. Cambridge: Cambridge University Press. (Kindle edition).

PART VIII

16

Making a Thing of It: Some Conceptual Commentary

DAVID PIMM

> Right now, it's only a notion. But I think I can get money to make it into a concept. And later turn it into an idea.
>
> (Paul Simon, as Tony Lacey in *Annie Hall*)

I am writing this chapter on the assumption that any readers it may have will have read many, perhaps all, of the other chapters that precede it in this book already, and that the former will consequently be able to identify the latter via the metonym of their authors' names. My intent in this concluding piece is to read slantwise across the chapters, rather than respond linearly to individual chapters in turn, as a book reviewer or book editors in an introduction might. My eye and ear will simply be out for things that catch my attention (whereupon I will blow a couple of them out of proportion[1]). For *things* lie at the heart of what this book is about, something I return to at the end of this chapter.

NAMING AND KNOWING

What are some of the mathematical concepts that are offered for examination and discussion in the chapters in this book? Even a quick scan reveals that *number* is present in all of the sixteen chapters, albeit to quite varying degrees. At the risk of sounding like an index entry, I found numbers finite and indefinite (Barany), infinite and many other named types (Netz), irrational and complex (de Freitas and Sinclair), imaginary, 'natural' (Corfield), rational, prime, *p*-adic (Harris), ordinal (Coles, in whose chapter number is almost everywhere), a cluster of these aforementioned types (Davis) and, lastly, Cantor's transfinite ordinal and 'paper' numbers (Nemirovsky).

[1] Mathematics educator Stephen Brown coined the transitive verb 'to bloop', an initialism arising from 'to blow [something] out of proportion'.

To be fair, Kennedy only mentions number once (unsurprising, given her focus), in a quotation from Netz about 'the number of diagrams', and only occasionally is there mention of number in Duffy ('real'), Plotnitsky ('infinite'), Mendick ('prime'), Barwell ('two-digit'), Brown ('number of things', *tout court*) and Roth (ditto). Nevertheless, as Netz observes, 'number' is the quintessential concept for historians of mathematics (and, as far as I am concerned, for others as well[2]). One striking thing arises simply from all these various things being called 'numbers', namely the way in which mathematical history eats itself by failing to mark conceptual change in this way. Same word, same thing! (For more on this, see Pimm, 2006, pp. 174–175.)

Corfield, in his chapter, makes some nice observations about options for mathematical baptism and the link between name and thing:

> With such little conceptual guidance provided by set theory, it is unsurprising that we turn to the naming of concepts for illumination, even if this may be misleading. It seems that there is a thin line between modifying the characterisation of a concept, and devising a new neighbouring concept. Sometimes we just add an epithet to an existing term, as in *imaginary* number or *nondifferentiable*[3] *function*. (p. 137; emphasis in original)

Concepts, then, are individuated, they have a topology ('neighbouring') and they are nouns (things).[4] In addition, such 'epithets' certainly echo through various names for 'types' of number: negative (positive), improper (vulgar (?), common (?), proper) irrational (rational), transcendental (algebraic), imaginary (real), complex, …

In the early seventeenth century, English mathematician Thomas Harriot dubbed complex roots 'noetical' (meaning 'arising from the mind'), though they could also, in present company, be termed 'conceptual numbers' (the Greek verb *noein* means 'to perceive'), but interestingly Harriot omitted them as solutions from his pedagogical writings, due to 'the preferences of his patrons' (see Lohne, 1979, pp. 296–298). Nemirovsky writes of Cantor's

[2] Henri Lebesgue was quite insistent that, even when examining older mathematical works, 'I don't do history of science, I do science' (quoted in May, 1966, p. 4). Lebesgue's comments cited in May's biographical sketch sit well alongside William Thurston's (1995) autobiographical remarks about two instances of his own mathematical activity. For more on Thurston's account, see below.

[3] The antonym of *nondifferentiable* is frequently the bourgeois (epithet?) 'well-behaved'.

[4] Linguist Michael Halliday (1993) has written extensively on nominalisation being a central force in the development of scientific thought, along with lexical density, interlocking definitions, syntactic ambiguity, grammatical metaphor and semantic discontinuity. In mathematics too, I believe, everything ends up as a noun and, as an aid to depersonalisation, as a noun in subject position (viz. concepts are seen to be animate, grammatically).

pejorative and marginalising term 'paper numbers' (for the reciprocals of transfinite ordinals).

Elsewhere (Pimm, 1988), I have observed how juxtaposing a novel adjective with an unsuspecting noun generally puts more pressure on the semantics of the noun, rather than on that of the adjective (e.g. spherical triangle, matrix multiplication,[5] differential geometry, etc.). It may seem unproblematic to assert (if based on nothing else than a knowledge of English) that a spherical triangle is a triangle. Certainly Bertrand Russell thought so:

> Before the time of Kant, it was generally held that whatever knowledge was *a priori* must be 'analytic'. What this word means will be best illustrated by examples. If I say, 'A bald man is a man', 'A plane figure is a figure', 'A bad poet is a poet', I make a purely analytic judgement: the subject spoken about is given as having at least two properties, of which one is singled out to be asserted of it. Such propositions as the above are trivial, and would never be enunciated in real life except by an orator preparing the way for a piece of sophistry. (1912/1970, p. 46)

Here goes! Such claims are only analytic in this sense *once a suitable concept modification has taken place* (a significant issue for pedagogy): in other words, they are contingent. And even then, especially pre-twentieth-century axiomatisation, there may be issues. Corfield also notes that Poincaré 'means' abelian group when he says 'group'. To which I might add that Galois wrote of *un groupement* (a grouping) rather than *un groupe*, a gerund which still carries a faint trace of the person doing the grouping. And his groupings were of a set of substitutions of roots of a polynomial equation.

But the history of the group concept during the nineteenth century is quite intricate, with the central notion being what later came to be called a binary operation on a set (Pimm, 1979). When a shift of attention came to 'groups of (space) transformations', Felix Klein (1873/1893), in his *Erlanger Programm*, observed, 'The idea and nomenclature are taken from the theory of substitutions, where one deals with the permutations of a finite number of discrete magnitudes, rather than the transformations of a continuous region' (p. 217). Both Klein and Sophus Lie[6] (e.g. 1895, as part of his eulogy for Galois) discussed whether an infinite group *must* contain an identity

[5] Tahta (1984) writes: There is a vague, often impenetrable, no-man's–land between the discourse of poets, philosophers or people in general and the discourse of mathematicians. People say things multiply when there is increase. Mathematicians also say they multiply when there is decrease (times half) or when neither increase nor decrease is in question (times a matrix). (p. 47)

[6] Part of Lie's own work involved trying to find a parallel to Galois theory for differential equations.

element, whether the operation *must* be associative, and whether all elements *must* have inverses. This is despite an 1860 entry in the English Cyclopaedia entitled 'Recent terminology in mathematics' (written by the English mathematician Arthur Cayley) asserting that *it* must have an identity.

In passing, Klein remarks that it was Lie who first pointed out to him that for groups of infinite order it was not a consequence of the group concept that the inverse of each element must appear in the group. Various axiomatisations of the concept of group took place in the US between 1902 and 1905, bringing about the possibility of the 'sophistic' proposition that 'a transformation group is a group', in certain instances. (This history has the potential for another Lakatosian account, by the way.)

I now have a far more general (if related) query. What does speaking of a 'mental object' (a commonplace description of a concept) do to the sense of 'object'? Recall Harriot's 'noetic' numbers, but also note Netz' use of the expression 'mental concepts' (p. 48). What other adjectives collocate with the term 'concept' and what pressure do they variously apply to it? In particular, what is the effect of 'mathematical' on 'mathematical concept'; what its effect on 'mathematical object'?

Lying beneath this question of concepts and their separateness or change is the vexed topic of metaphor, barely touched on explicitly in this book, although metaphors are used (e.g. Barnes' pyjamas in Barany's chapter and various ways of viewing concepts in the one by Davis). There are also two direct mentions that I noticed: Coles' brief allusion to the work of Jan Zwicky and her sense of the intellectual centrality of the act of 'seeing as' and de Freitas and Sinclair's negative mention in their first proposition: 'Concepts are not merely metaphors or representations' (p. 76).

For me, there is nothing 'mere' about metaphor, *especially* in the context of mathematics, though I understand their reaction to Lakoff and Núñez's (2000) account. Contrariwise, I certainly recall my strong, inner response of 'that's not right' upon hearing philosopher Chaim Perelman assert, 'There is no metaphor in mathematics' (in a 1976 talk at Cornell University). But perhaps there is not time nor room to get into this more deeply.

I will, however, end by simply offering psycholinguist Ellen Winner's (1979) assertion, that:

> In order to make possible a distinction between metaphorical and literal language, the definition of metaphor must be pushed one step further. It is not only the extension of a word to a novel referent on the ground of similarity that constitutes a metaphor: it is also the deliberate extension of this word ACROSS previously established category boundaries. (p. 472; emphasis in the original)

And observe that, in the case of 'number', this is something which has been done time and again, rather than come up with new names for new(-ish) things. So the notion of number must adjust and accommodate, as must its users, repeatedly. Mathematics is highly conservative in that regard, despite such conservatism causing significant pedagogic challenges by failing terminologically to mark the new.

How Big is a Concept?

One other consideration that reading this book brought up is the question of the scale of a concept (that of 'number' seems quite large, for instance) and my growing sense that pretty much everything now is spoken of as a concept (or sometimes as 'conceptual structure' or 'schema', a term introduced by Bartlett in the 1930s). Also, do concepts come individually or in clusters? Mathematical psychologist Richard Skemp (1979), in his slightly Victorian book *Intelligence, Learning and Action* (Victorian in the sense of attempting to offer a grand, general theory), worried at this (e.g. pp. 114–116), writing briefly on the *interiority* of a concept – a question of the degree of detail revealed when looking in, as if by means of a vari-focal lens.

In 1836, the writer Georg Büchner declared in *Woyzeck*, 'Every person is an abyss [*Abgrund*]. You get dizzy looking in.' Some forty years later – and contrasting it with Dedekind's notion of set which he denigrated as *einen geschlossenen Sack* ('a closed bag') – Cantor described his conception as follows: *Eine Menge stelle ich mir vor wie einen Abgrund* – 'I conceive of a set as an abyss.'[7] And the same might be said of concepts.[8]

Which brings me briefly to the question of what is meant by 'conceptual learning' (or sometimes 'concept learning'), occasionally offered as one of the polarising terms in public discussions of school mathematics, the other commonly called 'rote learning' (see Barwell and Abtahi's chapter in this book or the article by Chorney, Ng and Pimm, 2016). 'Conceptual' learning sounds like a type or form of learning, and 'rote' another, but what on earth do these terms mean? I realised in the course of writing this chapter that

[7] My translation: see Thiele (2005, p. 542).

[8] The way German concatenates nouns, prefixes and suffixes tempts me to coin a neologism (*die Abgrundlagen der Matematik*), given how often the term *die Grundlagen* shows up in mathematics – e.g. Nemirovsky writes at length about Cantor's book on sets entitled *Grundlagen einer allgemeinen Mannigfaltigkeitslehre* ('Foundations of a general theory of aggregates'), while Hilbert's *Grundlagen der Geometrie* is equally renowned. A gloss of *Abgrundlagen* might be 'abyss-foundations' or 'abyss-location', or perhaps, in the words of a tale told repeatedly by Thomas King (2003) in his Massey Lectures, 'Turtles all the way down'.

I don't know what it means 'to learn a concept'. I feel I know appreciably more what it means to learn *about* something or to learn *to do* something. In various of the chapters (most explicitly the one by Davis), the question of learning arose. And it is perhaps worth remembering that every mathematician was first a child.

A MATTER OF BIRTH AND DEATH

For me, the most pregnant overtones of the word 'concept' and 'conception' are, well, 'conceive' and 'conception'. Yup, the Latin roots agree: Latin *concipere* ('to take or hold together; to take in') and the related noun *conceptio* ('a becoming pregnant, a budding'). So the word 'concept' itself suggests it requires someone to give it birth. As Preston Hammer more than once remarked, 'the most neglected existence theorem in mathematics is the existence of people' (quoted in Brookes, 1970, p. vii).

In an informal course description, Canadian poet Lorna Crozier wrote, 'One of poetry's most ancient tasks is to remember, to mourn, to call the dead back to us in word and song'. And I'd like to juxtapose this with Kant's striking assertion that 'Mathematics is pure poetry', which could suggest that mathematics too has (or should have) an elegiac strand – it can be involved in calling back the dead (Bourbaki's *Éléments de Mathématique* as elegy?). As asked above, are historians of mathematics the authors of conceptual biographies? Or the poets of mathematical concepts?

Historian of science Catherine Chevalley, writing of her mathematician father Claude, commented:

> For him mathematical rigour consisted of producing a new object which could then become immutable. If you look at the way my father worked, it seems that it was this which counted more than anything, this production of an object which, subsequently, became inert, in short dead. It could no longer be altered or transformed. This was, however, without any negative connotation. Yet it should be said that my father was probably the only member of Bourbaki who saw mathematics as a means of putting objects to death for aesthetic reasons. (in Chouchan, 1995, pp. 37–38; my translation)

ON DRAWING AND DIAGRAMS, ON THEOREMS AND VISIONS

In some ways, the most singular chapter in this book is the one by Juliette Kennedy. First, she uses the word 'concept' only a handful of times (and the majority of occurrences are actually in quotations from others), most tellingly when she seems to wonder whether certain mathematical concepts are irreducibly visual. The second is with her constant focus on the visuo-gestural (as opposed to the verbo-graphical) in the service of *doing* mathematics, giving rise to my querying whether a diagram is/can be considered a concept. For the visual no less than the temporal has had a rough ride in the history of mathematics (see Pimm, 2006). And this despite the fact that, as René Thom (1971) notes, 'according to a long-forgotten etymology, a theorem is above all the object of a vision' (p. 697).

Kennedy, too, in the structure of her piece, enquires into the potential inseparability of syntax from semantics with regard to mathematical diagrams (which possibly relates to Yeats' questioning our ability – in his poem tellingly entitled 'Among school children' – to know the dancer from the dance). William Ivins, for thirty years curator of the department of prints at the New York Metropolitan Museum of Art and an extensive writer on images and objects and their connection to mathematics, framed it a little differently: 'Thus while there is very definitely a syntax in the putting together, the making, of visual images, once they are put together there is no syntax for the reading of their meaning' (1969, p. 61). For me, Ivins is drawing attention to a significant difference between 'the drawing' and 'the drawn', which have dramatically different relationships with time. And, as Wittgenstein (1953/1963) observed, 'Grammar tells what kind of object anything is' (p. 116).

Certainly the work of Gilles Châtelet, leant on in the chapter by de Freitas and Sinclair, is heavily focused on the human hand (and its traces) and its intellectual agency with respect to mathematical bringing-into-being.[9] Châtelet (1993/1999) wrote:

> They [diagrams] capture gestures mid-flight; [...] they are the moments where being is glimpsed smiling. Diagrams are in a degree the accomplices of poetic metaphor. (p. 10)

v IS FOR VECTOR

I start this section inspired by James Richardson's (2001) appropriately titled collection *Vectors: Aphorisms*[10] *& Ten-Second Essays*. One of his instances (out of more than several hundred such 'vectors' in his brief and elegant

[9] See also the excellent dissertation by Petra Menz (2015), where she examines the diagramming of research mathematicians and a doctoral student in a small seminar, not least the ways in which the diagrams themselves are attended to, are seemingly animate and, arguably, participate in the on-going mathematical conversation.

[10] The etymology of 'aphorism' is striking: at root, it comes from *aphorizein*, meaning 'to define'. Hippocrates (the doctor of the eponymous oath, not the mathematician who

book) runs in its entirety as follows: 'Greater than the temptations of beauty are those of method' (p. 26).

Beneath some of my sixteen attempts that I offer below (a format adopted in part due to fading space), however, I have added a paragraph or two of expansion. In the spirit of jazz, I call them fifty-second essays.[11]

1. The perfectoid space is an immaculate concept.
2. The term 'concept' is itself a metaphor, albeit a baby one.
3. Mathematical concepts are thought-fossils (mostly from extinct species?) See the Davis chapter.
4. What is the virtue of the virtual in mathematics?
5. Historians of mathematics write concept biographies; there are no concept autobiographies.

   Monk (2007), a significant biographer of both Wittgenstein and Russell, has written an extensive essay against attempts to provide biography with an underlying philosophical foundation. In particular, he reacts strongly against proposals from Freud onwards for psychoanalysis as providing such an underpinning. In contrast, the psychoanalyst Prophecy Coles (2003), in her work on siblings, tentatively proposes, 'psychoanalytic theory may be the theorization of autobiography' (p. 4). In relation to mathematics, the most subtle author I am aware of who wrote about these matters is Dick Tahta. In relation to this book (where Cantor and his ideas recur in several chapters), one of the final pieces Tahta (2007) wrote was entitled 'Recounting Cantor' (see also Pimm, 2007). There many things of interest here, not least his re-reading and revising of the two standard accounts (by Ivor Grattan-Guinness and Joseph Dauben), as well as a discussion of the theory of infinite sets being used to mathematise unconscious processes. Tahta's final sentence reads, 'A psychoanalytic account offers a possible way of reconciling the sharply different ontological views about mathematical objects[12] held by mathematicians and mathematics educators' (p. 10).

proved a quadrature of the lune) is often credited with the first set, starting with 'Art is long, life is short', although some parts of the Bible also share this poetic compression (e.g. Ecclesiastes – see later). But poet Elisa Gabbert (2015) claims aphorisms are more like assays – tests to run yourself up against – than definitions, as well as brief essays.

[11] This format also mirrors, to some extent, the propositional structure of de Freitas and Sinclair's chapter, though not all of mine are propositions and theirs are followed by ten-minute essays (see also Anne Carson's, 1992, *Short talks*). In addition, there is Zwicky's (2003) magnificent quotation-confronting aphoristic style deployed in *Wisdom & Metaphor* (mentioned in Coles' chapter).

[12] Tahta (2006) also wrote 'Sensible objects', a significant chapter on the nature of mathematical objects, replete with psychoanalytic overtones.

6. A mathematical concept is not a definition: but even definitions mutate (*mutatis mutandis*).

   One engaging source is Thurston (1995), wherein he documents eight different ways of conceiving of the derivative, although he numbers his final offering as #37 ('a sample entry from further down the list', p. 30). Each is of varying complexity and sophistication, without marking any one as definitive or the order as fixed. Thurston comments:

   > This is a list of different ways of *thinking about* or *conceiving of* the derivative, rather than a list of different *logical definitions*. Unless great efforts are made to maintain the tone and flavor of the original human insights, the differences start to evaporate as soon as the mental concepts are translated into precise, formal and explicit definitions. [...] The list continues: there is no reason for it ever to stop. (p. 30; italics in the original).

   Lakatos (1976) provides another such source, although his text is less biographical, more creative non-fiction. And, as mentioned above, throughout the nineteenth century, the concept of group shifted this way and that.

7. Mathematical concepts are rays.

   They have an origin but no end in sight. For instance, Lakatos' first footnote (1976, p. 6) mentions Euler's 'invention' of the concepts of vertex and edge (although see also Nemirovsky's closing image). Every concept has its history and at times a legal change of name – e.g. the names of various conic sections shifting due to Apollonius from *amblytome* (the obtuse-cut), *orthotome* (the right-cut) and *oxytome* (the acute-cut) to *hyperbola* (the exceeding in comparison), *parabola* (the comparable in comparison) and *ellipse* (the falling-short in comparison) – as well as having to contend with others apparently having the same name ('multivalued function', the Flemish composer Jacobus Clemens non Papa). What changes and what stays the same?

   Et nunc et semper. The desire for timelessness in mathematics (see the Chevalley quotation above and its link to aesthetics, as well as Pimm, 2006). Yet, as Zwicky (2003) tellingly asserts, 'The experience of beauty is the experience of some form (or other) of relief from time' (p. 71 LEFT).

   This suggests a concept's coming into being is more big bang than steady state. Conceptual conservation: concepts are neither created nor destroyed? But they do change form. Certainly not *Sicut erat in principio*.

8. Kant's *reine Dichtung* ('mathematics is pure poetry') and Bolzano's *Rein analytische Beweis …* ('Pure analytic proof …' – described in Kennedy).
   Wherein resides the purported purity of pure mathematics? See Mary Douglas' (1966) book *Purity and Danger.*

9. Lakatos and the emperor's new pyjamas.

   Barany's evoking of Barnes' pyjama metaphor draws important attention to labels and the human act of labelling (and not the labels *in* the pyjamas, which may prove to be the grammatical entry in the lexicon) – one which is strikingly exemplified in Roth's example of young children learning to issue classifications *geometrically.* Barany also connects labelling to mathematical practice – not the overt focus of this book, but one which nonetheless has a strong shadow existence throughout (not least in Kennedy's chapter) and is often in contrast to discussion of concepts.

   And why does Lakatos crop up in six different chapters? But he may not be quite the same 'Lakatos' to every author in the book – a list which includes, in no particular order, mathematics educators, historians of mathematics, mathematicians and philosophers of mathematics. Will the real Imre Lakatos please stand up (see Pimm, Beiseigel & Meglis, 2008)?

10. Number of numbers, saith the Cantor, all is number (with apologies to Ecclesiastes).

    (Except for geometry and its discontents, perhaps – those diagrams.) From number as answer to the question 'how many?' to number as root of an equation (itself an idea that shifts dramatically over time) to number as determined by the company it keeps (e.g. number fields or number domains, both of which have technical senses).[13] With regard to the future, see Sinclair's (2015) chapter, 'What kind of thing might number become?'.

11. What do cuboid (Roth), groupoid (Corfield) and perfectoid (Harris) have in common?

    The '-oid' suffix (often rendered as '-like') can signal a conceptual blurring while evoking a similarity, without necessarily being the thing itself: asteroid, opioid, …. It relates to the Greek word *eidos*, meaning 'form'. The age-old inclusion/exclusion debate: is a cube a cuboid, is a

[13] The name 'Ecclesiastes', by the way, is a transliteration from Hebrew, a word that means 'One who convenes or addresses an assembly' – see shortly the discussion of the etymology of 'thing'. And even Richard Dawkins (see Davis' chapter) is a fan of Ecclesiastes: www.theguardian.com/science/2012/may/19/richard-dawkins-king-james-bible

group a groupoid?[14] Are these indeed the conceptual prototypes? But clearly a perfectoid space is not (quite) perfect? Or possibly more general than perfect? Pluperfect?

Worse, while an element of a vector space is a vector, and an element of a function space is a function, an element of a perfectoid space is not a perfectoid. (Harris' title suggests 'perfectoid' as the concept, but 'perfectoid space' is actually the concept in question.) It sounds like a last-minute hedge, a perfect-like space rather than a perfect one? Yet, nonetheless there is a mapping called 'perfection' from {Algebraic varieties/$K^{\flat}$} to {perfectoid spaces/$K^{\flat}$} (see Bhatt, 2014, p. 1084).

12. Is 'almost mathematics' a hedge?

    There are very few hedges in formal written mathematics: 'almost everywhere' sounds like one, but has quite a specific meaning.

13. A goodly number of the mathematicians mentioned are nineteenth-century German-speakers, while many of the philosophers drawn on are twentieth-century French-speakers.

    Bolzano, Cantor, Dedekind, Frege (?), Riemann, ...
    Althusser, Châtelet, Deleuze, Derrida, Foucault, Lacan, ...
    Why not the other way round?

14. Euclidean diagrams are about essence, not existence (Lachterman, 1989).

15. What is the opposite of reification? Or is it always a one-way street?

    Phenomenology is one name for the path back from the object to the thing, the counterbalance to objectification, or 'progress'. Poetry is another. (McKay, 2012, p. 66; italics in the original)

16. Mathematical concepts are life forms: the concept is long, human life is short.

MATOI

An object is a thing under surveillance.
(McKay, 2012, p. 55; italics in the original)

Just as my opening quotation from *Annie Hall* suggests that notion, concept and thing are distinguishable (and *pace* Davis' footnote 1), I would like to propose another related trilogy for consideration: concept, object and thing. In English idiom, 'making a thing of it' involves the conscious

[14] Wolfram Mathworld documents three quite different definitions of groupoid: see http://mathworld.wolfram.com/Groupoid.html

expansion of the purported significance of something – of some thing. But in its second, more literal sense, equally pertinent here I feel (namely making something into a thing), it seems the 'it' has already been objectified by the English language (via the very use of the pronoun 'it'). Surprisingly perhaps, there is also a third sense for Matoi, where the sentence stress is on the final word, which, drawing on McKay's distinction indicated in his pithy aphorism above and discussed further below, suggests the possibility of going back from an object to a thing (or in the case of mathematics possibly going forward from a mathematical object to a mathematical thing, something it has possibly never yet been).[15]

The essence of a striking lecture/reading Don McKay (2011) gave at the Vancouver Readers and Writers Festival involved distinguishing 'object' from 'thing'. In it, he detailed how language does a fair bit of the work of our human need to objectify the world (in the sense of creating temporally and spatially stable objects from the flux – a.k.a. nouns – which I contrast with Smaling's intriguing notion of objectivity viewed as 'doing justice to the object'; see van den Brink, 1990). And McKay would like to establish a return ferry route from object back to thing, which he links to the etymology of the word 'metaphor' itself – *meta+pherein* ('to carry across'). For McKay, *thing* (including its Old English and Icelandic etymological roots, meaning 'assembly', 'gathering', 'parliament' – see footnote 13)[16] has a greater sense of self-possession and independent existence from us (and from our overbearing ways of seeing) than *object*: so much so, that some of his 'thing-poems' are second-personned, where he addresses the beloved things as 'you'.

Martin Buber (1923/1958), in his book *I and Thou*, discusses two sorts of significant human connections: I–It, which he terms 'experiences', and I–Thou, which he calls 'relations' (cf. Coles' chapter). His book is very much concerned with intimacy and he discusses how Thous can become Its when the relationship changes (e.g. through neglect or being forgotten[17]). But Its can become Thous when suddenly recognised in a certain way and then live an active part in one's life, arising from and resulting in a mutual

[15] Douglas Hofstadter (1997), writing a quasi-autobiographical account of his unexpected involvement with Euclidean geometry, at the bottom of p. 8 first uses the possessive pronoun 'my' in relation to a line segment (and then three points on it and then 'my crystal') that he has come up with ('discovered' is his term). Later on in his account, the possessiveness disappears: object, not thing.

[16] Later in life, my parents lived on the Wirral peninsula in the UK, near a village called Thingwall, meaning 'assembly field'.

[17] Sociologist Charles Fisher (1966, 1967) has written an account of what he terms the 'death' of a mathematical theory (nineteenth-century invariant theory), an account that can be seen in terms of concepts moving back from Thou to It.

recognition that both are there. Now, Buber was mostly concerned with the animate, but if mathematical concepts can be seen as life forms, especially to their progenitors, then ….

Buber also wrote of the *Eswelt* (the It-world)[18] and the *Dingwelt* (the Thing-world). He observes:

> *ohne Es kann der Mensch nicht leben. Aber wer mit ihm allein lebt, ist nicht der Mensch.* (p. 44)
>
> (without It one cannot live. But whoever lives with It alone is not human.)

I end here, still wondering about links among Buber's *Eswelt*, McKay's distinction between object and thing, and the world of mathematics: how certain concepts for some become worthy of Thou status (grammatically moving from 'that' or 'which' to 'who') and how, especially for mathematics learners, such a more intimate relation might be engendered.

REFERENCES

Bhatt, B. (2014). What is a perfectoid space? *Notices of the AMS*, 61(9), 1082–1084.

Brookes, W. (1970). Preface. In ATM members (Eds), *Mathematical reflections: Contributions to mathematical thought and teaching, written in memory of A.G. Sillitto* (p. vii). Cambridge: Cambridge University Press.

Buber, M. (1923/1958). *I and Thou* (R. Smith, trans.). New York: Charles Scribner's Sons.

Carson, A. (1992). *Short talks*. London, ON: Brick Books.

Châtelet, G. (1993/2000). *Figuring space: Philosophy, mathematics and physics*. Dordrecht, The Netherlands: Kluwer Academic Publishers.

Chorney, S., Ng, O. & Pimm, D. (2016). A tale of two more metaphors: Storylines about mathematics education in Canadian national media. *Canadian Journal of Science, Mathematics and Technology Education*, 16(4), 402–418.

Chouchan, M. (1995). *Nicolas Bourbaki: Faits et légendes*. Argenteuil, France: Éditions du Choix.

Coles, P. (2003). *The importance of sibling relationships in psychoanalysis*. London: Karnac Books.

Douglas, M. (1966). *Purity and danger: An analysis of concepts of pollution and taboo*. London: Routledge and Kegan Paul.

Fisher, C. (1966). The death of a mathematical theory: A study in the sociology of knowledge. *Archive for History of Exact Sciences*, 3(2), 137–159.

(1967). The last invariant theorists: A sociological study of the collective biographies of mathematical specialists. *European Journal of Sociology*, 8(2), 216–244.

Gabbert, E. (2015). Aphorisms are essays. *The Smart Set*. Retrieved from: http://thesmartset.com/aphorisms-as-essays/

[18] And given the English translation of Freud's direct and plain German terminology (*Das Ich, Das Über-ich* and *das Es*) in terms of James and Alix Strachey's Latinate (and hence distancing) rendering of these terms (as the ego, the superego and the id), the *Eswelt* is also the id-world.

Halliday, M. (1993). Some grammatical problems in scientific English. In M. Halliday & J. Martin (Eds), *Writing science: Literacy and discursive power* (pp. 69–85). London: The Falmer Press.

Hofstadter, D. (1997). Discovery and dissection of a geometric gem. In J. King and D. Schattschneider (Eds), *Geometry turned on: Dynamic software in learning, teaching, and research* (pp. 3–14). Washington, DC: The Mathematical Association of America.

Ivins, W. (1969). *Prints and visual communication*. New York: Da Capo Press.

King, T. (2003). *The truth about stories: A native narrative*. Toronto, ON: House of Anansi Press.

Klein, F. (1873/1893). A comparative review of recent researches in geometry. *Bulletin of the New York Mathematical Society*, 2(10), 215–249.

Lachterman, D. (1989). *The ethics of geometry: A genealogy of modernity*. London: Routledge.

Lakatos, I. (1976). *Proofs and refutations: The logic of mathematical discovery*. Cambridge: Cambridge University Press.

Lakoff, G. & Núñez, R. (2000). *Where mathematics comes from: How the embodied mind brings mathematics into being*. New York: Basic Books.

Lie, S. (1895). Influence de Galois sur le développement des mathématiques. In *Le Centenaire de l'École Normale: 1795–1895* (pp. 481–489). Paris: Librairie Hachette et Cie.

Lohne, J. (1979). Essays on Thomas Harriot. *Archive for History of Exact Sciences*, 20(3), 189–312.

May, K. (1966). Biographical sketch of Henri Lebesgue. In K. May (Ed.), *Henri Lebesgue: Measure and the integral* (pp. 1–7). San Francisco, CA: Holden-Day.

McKay, D. (2011). 'We're not the centre of the universe'. Talk given on October 23, Vancouver, BC: The Vancouver Writers Festival. Retrieved from: www.writersfest.bc.ca/audio-archives/were-not-centre-universe-don-mckay

(2012). *Paradoxides*. Toronto, ON: McClelland & Stewart.

Menz, P. (2015). *Unfolding of diagramming and gesturing between mathematics graduate student and supervisor during research meetings*. Unpublished Ph.D. dissertation, Burnaby, BC: Faculty of Mathematics, Simon Fraser University.

Monk, R. (2007). Life without theory: Biography as an exemplar of philosophical understanding. *Poetics Today*, 28(3), 527–570.

Pimm, D. (1979). *Klein, Lie and the changing concept of group*. Unpublished Master's essay, Madison, WI: Dept of History of Science, University of Madison–Wisconsin.

(1988). Mathematical metaphor. *For the Learning of Mathematics*, 8(1), 30–34.

(2006). Drawing on the image in mathematics and art. In N. Sinclair, D. Pimm & W. Higginson (Eds), *Mathematics and the aesthetic: New approaches to an ancient affinity* (pp. 160–189). New York: Springer.

(2007). Afterwords. *For the Learning of Mathematics*, 27(3), 24.

Pimm, D., Beiseigel, M. & Meglis, I. (2008). Will the real Imre Lakatos please stand up? *Interchange*, 39(4), 467–478.

Richardson, J. (2001). *Vectors: Aphorisms & ten-second essays*. Keene, NY: Ausable Press.

Russell, B. (1912/1970). *The problems of philosophy*. Oxford: Oxford University Press.

Sinclair, N. (2015). What kind of thing might number become? In E. Davis & P. Davis (Eds), *Mathematics, substance and surmise: Views on the meaning and ontology of mathematics* (pp. 347–363). New York: Springer.
Skemp, R. (1979). *Intelligence, learning, and action: A foundation for theory and practice in education*. Chichester, UK: John Wiley & Sons.
Tahta, D. (1984). Book review of R. Rucker, *Infinity and the mind*. *Mathematics Teaching*, 107, 47.
(2006). Sensible objects. In N. Sinclair, D. Pimm & W. Higginson (Eds), *Mathematics and the aesthetic: New approaches to an ancient affinity* (pp. 191–222). New York: Springer.
(2007). Recounting Cantor. *For the Learning of Mathematics*, 27(3), 8–11.
Thiele, R. (2005). Georg Cantor (1845–1918). In T. Koetsier & L. Bergmans (Eds), *Mathematics and the divine: A historical study* (pp. 523–548). Amsterdam, The Netherlands: Elsevier.
Thom, R. (1971). 'Modern mathematics': An educational and philosophic error? *American Scientist*, 59(6), 695–699.
Thurston, W. (1995). On proof and progress in mathematics. *For the Learning of Mathematics*, 15(1), 29–37.
van den Brink, J. (1990). Classroom research. *For the Learning of Mathematics*, 10(1), 35–38.
Winner, E. (1979). New names for old things: The emergence of metaphoric language. *Journal for Child Language*, 6(3), 469–491.
Wittgenstein, L. (1953/1963, 2nd edn). *Philosophical investigations* (G. Anscombe, trans.). Oxford: Blackwell.
Zwicky, J. (2003). *Wisdom & metaphor*. Kentville, NS: Gaspereau Press.

INDEX